AF522207

Encyclopaedia of
PLANT ECOLOGY

Encyclopaedia of
PLANT ECOLOGY

Volume 3

Shehzad Ahmad

ANMOL PUBLICATIONS PVT. LTD.
NEW DELHI - 110 002 (INDIA)

ANMOL PUBLICATIONS PVT. LTD.
H.O.: 4374/4B, Ansari Road, Darya Ganj,
New Delhi-110 002 (India)
Ph.: 23278000, 23261597
B.O.: No. 1015, Ist Main Road, BSK IIIrd Stage
IIIrd Phase, IIIrd Block,
Bangalore - 560 085 (India)
Visit us at: www.anmolpublications.com

Encyclopaedia of Plant Ecology

First Published, 2007

ISBN 978-81-261-3398-7 (Set)

PRINTED IN INDIA

Printed at Amit Enterprises

Contents

Preface

Plant Ecology is a newer sub-discipline of Ecology. Plant Ecology is the study of plants and trees etc. in their environment, especially how they relate to each other. In fact, all the living beings including plants and their direct or indirect relation to the non-livings, constitute the ecosystem. For a healthy life, they all play vital roles and for a sound ecosystem, a few precautions, taken by us, are necessary.

Environment is the most significant factor for the existence of living beings, including plants. It includes climate, pollution, deforestation, conservation, disasters of flora and fauna and natural resources, etc. Not to say that, the present condition of the ecosystem is in a sorry state. With decreasing forests, rapid growth of industrialisation and globalisation etc., have left, in its wake, a number of other problems, challenging and threatening the whole existence of the vegetation and greenery on earth. All these issues demand our urgent heed. They have caused a huge loss to the natural balance. If we do not put a check on them, then different crises, looming large would become inevitable and incurable.

The prevailing condition makes it incumbent upon every one of us to try our level best to think over the ways; how to minimise the effect of these fatal factors. We should also be considerate enough to preach others, as well, to take the same initiative; components of ecosystem, *i.e.*, plants and animals could be saved from destruction for now and in future.

This work, *"Encyclopaedia of Plant Ecology"* is a precious gift to every one concerned with the subject. Prepared with utmost care, it would, undoubtedly, serve a great purpose.

All wise and positive comments from scholars, professionals and readers are welcome.

Editor

1

ECOLOGICAL SAFETY

SPECIFIC DIMENSIONS

(i) Dust and smoke in the air do not allow a clear view of nature's beauty.

(ii) Foul odours emitted by industries, automobiles, dirty drains and garbage heaps in cities are a great nuisance.

(iii) Gases like oxides of sulphur and nitrogen alongwith smoke make breathing difficult.

(iv) The stone of Pantheons in Athens has deteriorated in the past 50 years from air pollution.

(v) Similarly statue of liberty is corroded from SO_2 and NO_2 and Taj Mahal from SO_2 emitted by Mathura refineries.

RISK RELATED TO ATMOSPHERE

Carbon dioxide content of the air is increasing due to deforestation and combustion in industries, automobiles and planes and is likely to become double by 2020. This increase is affecting the atmospheric composition and balance of gases, which are among the factors that controls earth climate. Increase of CO, may cause rise in atmospheric temperature producing what is called greenhouse effect. Aerosols deplete the ozone layer in the stratosphere. Thinning of ozone layer would permit more of the harmful ultraviolet rays to reach the earth and may cause extensive damage to plants as well as animals.

Specific Features of Earth

Land is the solid, exposed surface of the earth district from oceans, lakes, etc.Land forms about one-fifth of the earth's surface. It measures about 13,939 million hectares. About 36.6 per cent of the land area is covered by houses factories, roads, railways, deserts and dunes, glaciers, mountains and polar ice marshes, about 30 per cent by forests, and about 22 per cent by meadows and pastures. Only 11 per cent of land area is fit for tilling.

Upper Layer of Environment

It is the top fertile layer of the earth capable of supporting plant growth. It is a dynamic layer in which many chemical, physical and biological activities are going on constantly. It converse about four fifth of the land area. "Soil is the most precious thing of nation, region of locality. It is the base on which the human civilization thrives. Among the boundaries of 'NATURE' none is more important for the survival of the human race than the soil."

The scientific study of soil is called pedology. It deals with the origin, formation and geographic distribution of the soil. Soil consists of five components (i) mineral matter (ii) organic matter (iii) water (iv) air and (v) living organisms. All these components are essential for proper plant growth.

Mineral Matter: Soil is an aggregation of mineral particles of different size and shapes. The assortment of these particles differ from one soil type to another. The mineral particles are derived from the under lying parent rock by its weathering or disintegration. Weathering of rocks involve physical and chemical break down. The former occurs to temperature variations, alternate drying and wetting, microbial activities, action of plant roots and burrowing animals while latter occurs due to oxidation, reduction, hydration, carbonation, etc.

The mineral matter in the soil occur as particles of different size and shape. Depending upon size 5 types of mineral particles are recognised gravel (2-50mm2). coarse sand (2-0.2 mm), fine sand (0.2-0.02mm), silt (0.02-0.002 mm) and clay (less than 0.002mm). While gravel and sand particles are visible with naked eyes, the silt particles can be seen with the help of light microscope and the clay by an electron microscope. Gravel consists of stones. Sand and silt consists largely of quartz (silicon dioxide, SiO_2) and are chemically inert. Clay contains mineral salts is chemically active and has great water holding capacity. Soil texture spends upon the proportion in the which the various types of particles are mixed. The best soil for the growth of the plants loamy soil. A good loamy soil contains 1 part clay, 2 parts silt and 2 parts sand. A clay soil is not good for plant growth because it is impermeable to water. The sand soil is also not good because it can not hold water.

Transported Mineral Particles

(a) Soil transported by gravity are called alluvial and form mud flows and land slides, etc. Most alluvial deposits on the western Himalayas on stabilisation bear chirpine community.

(b) The soil transported by running water are called alluvial. These occur on out wash plains, flood plains, terraces, etc. The flood plain deposits in Himalayas bear chirpine, deodar, etc. In plains Acacia, Delbergia, etc. present.

(c) Wind blown deposits bear sand which gradually deposits in the form of an arc. It has salt binder plants like prosopis.

Organic Matter: It is formed in the soil by decay of dead plant pasts, animal wastes and dead animals, which get decomposed by the action of bacteria and fungi present in the soil. The mineral matter alone called parent earth. They become soil when mixed with organic matter.

Organic matter is thoroughly mixed with the mineral matter by burrowing animals like earthworms, centipedes, millipedes, insects, etc. The organic matter gives brown colour to the soil. The organic matter of the soil form humus. It is brown in colour, spongy in texture. When present in clay soil it looses the soil particles and make it porous but in sand soil it acts as weak cement to hold the sand particles together so that water holding capacity is increased.

Water: The space between soil particles is filled with water or the source of water in the soil is rain water. After heavy rain most of the water drains away along the slopes called run away water. Some of the water enter in the soil and moves down due to force of gravity called gravitational water which finally reaches the underground water table. Some of the water retained by the soil particles against force of gravity called field capacity or water holding capacity of the soil. Some of the water forms field capacity remains tightly absorbed on the surface of soil particles and is not absorbed by the plants called hygroscopic water. Remaining water fills the interspaces of the soil particles called capillary water. It is the only water which is all the time available to the plants.

Air: The interspaces of the soil particles also contain air. It is inversely proportional to the amount of water. The best soil for the growth of plant is that which interspaces have water and air in equal proportion. Clayey soil is poorly aerated as it contains very tiny pore spaces. This makes it unsuitable for plant growth which require oxygen.

Living Organism: A variety of organisms live in the soil. These include flora (plant) as well as fauna (animals)

(i) Microfauna, i.e, protozoans visible with microscope. These include Amoeba, Euglena, paramecium, etc.

(ii) Macrofauna, *.i.e.*, animals visible with naked eyes. These include earthworms, centipedes, millipedes, insects, termites, snails, rats, etc.

(iii) Microflora, comprising bacteria such as Rhizobium, Azobacter, Clostridium, etc.

(iv) Microflora, includes algae, moulds, mushrooms, etc.

Fertility of Soil

Soil profile shows 4 distinct regions called horizons. Horizon A in the top soil. It is darker and of a looser texture than the underlying horizon B. Plant and animal matter collects at the surface of this horizon, forming the litter. Below the litter is the humus, *.i.e.*, organic matter undergoing decay by microbial-action. The rest of horizon is rich in organic and mineral contents. The horizon B has soil particles smaller and usually more compacted than in the horizon A. Minerals brought by rain water from the upper horizon collect in this horizon. The horizon C consists of weathered materials derived from the intact parent rock that forms the horizon D.

Influence on Soil

Soil is the major and only plant life sustaining resource on earth which continues the flow of energy on the earth. But over exploitation and improper use of this resource by man leads to its depletion. This degradation of soil is mainly due to two factors. Soil erosion and soil pollution.

Soil Erosion: The removal of top fertile soil from its resting place by various physical agencies like wind, water is called soil erosion.

Water Erosion: It is the soil erosion caused by the agency of water. Water erosion is maximum during melting of snow and heavy rainfall which can not be absorbed by soil. Soil cover and slop of the area determine the degree of erosion. Heavy rain fall directly bombard the soil and churn up the same. The compaction caused by falling rain drops and water borne soil particles clog all the soil pores. There is no more absorption of water and it collects on the surface of the soil which moves along with slopes. It is called run off which takes away suspended soil particles and causes soil erosion. Depending upon the form of the lost soil it may be:

Sheet Erosion: Here the removed soil is like a thin covering from large area. This sheet is lost more or less uniformly. It occurs on smooth and gentle slopes. Sheet erosion causes thinning of surface layers of the soil. It gives rise to areas of light colour or galled spots.

Rill Erosion: If sheet erosion occurs with full force, the run off water moves rapidly over the soil surface cutting well defined finger shaped groove like structures, appearing as thin channels or streams called rills. The rills functions as narrow water channel in which flowing water picks up more speed and higher cutting power.

Gully Erosion: This results due to the convergence of several rills (thin channels formed during rill erosion) towards the steep slope, which form together wider channels (grooves) of water, known as gullies. Gullies are either V or U shaped depending upon width they are designated as small and large. Gullies not only cause the loss of land to the formers but also cut the field into fragments which can not be ploughed or harvested together.

Wind Erosion: It is the removal of top fertile soil through the agency of wind. Soil erosion by wind is common in dry (arid) regions where soil is chiefly sandy and the vegetation is very poor or even absent. In our country wind erosion affects about 50 million hectares of land, most of which is in Rajasthan. As in water erosion, wind erosion also triggered by the destruction of natural vegetation cover of land by over-felling and overgrazing. Once the top soil is laid bane to the fury of strong gales, it gets blown off in the form of dust storms and sand storm. Depending upon the whole mechanisms of the soil removal, this may be of the following types.

(a) *Suspension:* The fine soil particles the size of less than 1mm., get suspended in wind. They are carried as dust.

The dust storms contain these particles in large number. These particles are deposited several kilometres away when the wind velocity decreases.

(b) *Sanitation:* In such arid regions where rainfall is low, drainage is poor, and high temperature prevails, water evaporates quickly leaving behind the salts. Salts are normally chlorides, sulphates, carbonates and nitrates of potassium, magnesium and sodium and chlorides and nitrates of calcium. The major portion of such salty soil is carried by wind in the form of small leaps which is caused by direct pressure of wind on small particles of soil.

(c) *Surface Creep:* The heavier particles of soil (5-10 mm) that are not easily thrown up by wind are simply pushed along the ground by the striking of the smaller particles. Surface creep is therefore, an indirect movement.

Wind removes away the fine particles from the exposed soils. Only coarser particles of the size of sand remain behind. The process of wind erosion, therefore, makes the soil sandy. The sandy soil is neither chemically fertile nor capable of retaining water. It therefore, becomes highly unstable and liable to shift with the movements of wind. This gives rise to sand dunes which are even more unstable.

Removal or top soil exposes the root system of the plants growing in that soil. They get undergo desiccation and get killed. Many plants get burned inside sand dunes. On settling, wind borne sand particles damage irrigation channels, ponds, lakes and fertile soil. The deposition of sand particles on fertile soil makes the latter barren. This causes the spread of deserts in the direction of wind.

Effects of Soil Erosion

(1) Soil erosion causes the removal of top fertile soil. The soil exposed after erosion is less fertile as a result vegetation cover of the soil is reduced.

(2) Erosion on the hill slopes destroy the forest vegetation of the mountain and foot hills.

(3) With the lost of forest vegetation due to soil erosion, the wild life of the area also gets destroyed.

(4) Thinning of soil layer reduces the plant growth, minerals and water.

(5) Crop production is also reduced.

(6) Total rain fall of the area is reduced because of less vegetation due to soil erosion.

(7) In the absence of vegetation as rain water is not readily absorbed, it rapidly passes along the slopes. It produces flash floods in plains.

(8) Soil erosion cause disturbance in air, humidity balance.

(9) Soil erosion leads to reduced productivity of the land, reduced water availability for irrigation and reduced hydroelectric power during dry period of the year. Hence famine overtakes the area.

(10) Soil erosion causes the development of deserts.

Mechanic Skills of Preservation

Soil conservation is the maintenance of soil fertility by preventing soil erosion, protection against fire, waste and misuse, planning judicious use and use of fertilizers. Due to lack of proper conservation methods the vast areas of fertile land have been converted into wastelands. There are several ways of conserving soil.

Biological Methods: They are meant for keeping the soil under cover for maximum period of time. Soil cover, especially, that of living plants, decreases water erosion. Erosion becomes negligible under continuous plant cover. Biological methods are of three types—agronomic, dry farming and agrostological.

Agronomic Methods: Natural protection by growing vegetation in a manner that reduces soil loss. These are:

(i) *Addition of Fertilizers:* Density of plant growth is

dependent on the fertility, hydration and aeration of the soil. Soil fertility is maintained by the addition of manure and fertilizers. It improves both aeration and hydration. Presence of manure decreases run off.

(ii) *Crop Rotation:* It is the practice of growing different crop plants in successive years on the some piece of land. It decreases soil loss and preserves the productivity of land. The same crop year after year depletes the soil minerals. This is overcome by clouting legumes.

(iii) *Mixed Cropping:* Two or more crops are grown simultaneously on the same piece of land, *e.g.* Millet, Black gram and pigeonpea the method avoids the risk of crop failure, provides a better use of soil fertility and check soil erosion.

(iv) *Mulching:* It is effective against wind as well as water erosion. Soil is allowed to remain untilled. It is covered with grasses, straw, leaves, crop, residue and other form of plant litter. Mulches (2-3" thick) reduce soil moisture by addition of organic matter to soil. The covered soil does not come in direct contact with the agencies of erosion.

(v) *Contour Farming:* It is normally performed on slopes. The land is ploughed at right angles to the direction of slope. It produces alternate furrows and ridges around the slope. Ridges at the same level are known as 'contours' the water is caught and held in furrows and stored, which reduces run off and erosion.

(vi) *Strip Cropping:* It involves the planting of crop in rows or strips to check flow of water. It may be contour strip cropping (strips planted along the contour at 90° to the direction of slope), field strip cropping (strip planted parallel to each others), or wind strip cropping (strip planted in straight paralleled rows at 90° to the direction of prevailing wind.)

(vii) *Fallowing:* It is an old method to improve soil fertility and prevent soil erosion. After harvesting of a crop, the land is left untilled for one or more season. But the practice is not possible modern-day India when the pressure on the land is already severe.

Dry Farming: This practice is useful for croplands grown in low and moderate rainfall areas, where ordinary farming is at risk, crop production, animal husbandry and growing grazing fields are the only possibilities of checking erosion. Method employed differ in different areas. Some of them are fallowing the land, strip cropping, crop rotation, contour farming, etc.

Agrostological Methods

(i) *Retiring the Land:* The land is taken out of cultivation permanently in areas which are prone to heavy erosion, especially on the sloping mountains. The land is ploughed and sown with grasses. Grazing is not allowed in first year controlled grazing is permitted after soil erosion has stopped and the soil becomes stabilised. The grasses use to check soil erosion are cyanodon dactylon, and dactylis glomerata.

(ii) *Lay Farming:* This aims at to grow grasses in rotation with field crops like Jowar and Gingelly (Sesamum indicum), which help in building up the structure of soil, preventing soil erosion and improving its fertility.

(iii) *Controlled Grazing:* Excessive grazing makes a land baren and the soil particles exposed become dry. The underground roots become weak due to lack of foliage and the soil becomes dry and weak which is easily eroded by wind and water.

(iv) *Afforestation and Reforestation:* Afforestation is the formation of forest where no forests existed previously. Reforestation is the replantation of forests which have been previously destroyed by felling, fire or over-grazing. In hilly areas deforestation results in reduced frequency of rain fall, increased melting of snow, formation of temporary rivulets in rainy season, occurrence of floods and damage to agriculture and property. Afforestation and Reforestation help to check soil erosion and therefore, the loss occurred through it.

Mechanical Methods: These methods are used as supplements to biological methods. These methods help in increasing water retentivity of the soil, decrease the velocity of run off and prevent soil erosion. These are:

(i) *Basin Listing:* In this a number of small basin or furrows are constructed along the slopes or contour to retain water which also reduces its velocity.

(ii) *Levelling:* Construction of small gullies, grooves and undulation in the path of the slop increase chances of erosion. These are filled up and soil is levelled so that water absorption increases and erosion is prevented.

(iii) *Terraces:* It means the division of sloppy areas into series of small flat fields called terraces by mean of ridges. Because of their step like appearance they are called bench terraces. The terraces slow down the velocity of run off and check erosion.

(iv) *Ridge Terracing:* These are constructed in those areas where terraces cannot be built. Here small ridges are constructed at a distance of 1 -2m throughout the slope for retaining moisture and prevention of soil erosion.

(v) *Gully and Ravine Control:* To check the formation of widening of gullies by constructing bunds, dams, drains or diversions through which excess run off water channelled. The check dams reduce the rapidity of water flow and are slowly filled up with the deposition of silt.

(vi) *Construction of Dams:* Floods are the regular- feature of every year in the rivers. Floods are the major causes of soil erosion. This wide spread erosion can be checked by constructing dams. These dams provide water for irrigation and unfertile lands can be converted to fertile ones.

Unfortunately, so far very little effective work has been done in India to combat wind and water erosion. It is estimated that a programme for the control of wind erosion covering 50 million hectares would cost 3000 crores of rupees. This is assuming an average cost of not more than Rs. 600 per hectare to carry out necessary afforestation, grassing and protective measures.

So far roughly about Rs. 300 crores have been spent on soil conservation since the beginning of the plan more than 20 years ago.

Green House Effect

Raising of temperature by allowing solar radiations to pass in but preventing long wave heat radiations to pass out is termed aMd respiration. It is estimated that more than 18 x 10^{12} tonnes of CO_2 is being produced annually from the fossil fuels only. Not only CO_2, CFCs and oxides of nitrogen and methane also exert green house effect.

CO_2 concentration has increased from 280 ppm in 1800 to 359 ppm in 1994 due to increasing use of fossil fuels and decreasing forest cover. More methane is being released from paddy fields due to intensive cultivation. Combustion of fossil fuels is also the causative agent for higher concentration of nitrogen oxides. The react with hydrocarbons to form ozone. Higher concentration of green house gases is slowly warming up the earth.

1. Global warming shall cause partial melting of polar and alpine ice caps that would result in raising sea level. Already the sea level has risen 15cm in the past century. If the trend continues, there is a danger of submersion of large areas, *e.g.*, Maldives and six other coral at all countries, several thousand islands.
2. Grain production will be reduced.
3. There will be a shift in rainfall and climatic zones.
4. Deserts are likely to increase.

5. Chances of hurricanes, cyclones and floods will be more.
6. The forests present in the middle latitudes will be wiped out.
7. Several lakes would dry up.

So, UNEP has chosen the slogan "Global warming: Global warming": and since 1989, 5th June is celebrated as World Environment Day. The cost of defence (reduction of gas emissions and research to identify the hardest hit regions and plan of coastal defence) would be enormous.

Ozone layer is present in the stratosphere (15-50 km height). It is concentrated at a height of 20-25 km. Ozone is responsible for protecting the earth from high energy Ultraviolet radiations by changing the same infra red rays. It forms a life saving screen as it checks the entry of lethal UV-rays. The depletion of this O_3 layer by human activities may have serious implications and this has become a subject of much concern over the last few years.

The first horrifying report of alarming ozone depletion came from British Scientists who reported an Ozone hole in the Ozone layer over Antartica in 1985. This, Ozone hole, has grown in size over the years. In 1994 alone, it widened from 129-133 cloboson units. A similar but smaller hole has also appeared over North Pole. Size of the holes varies with the season. Damage has occurred to the ozone layer at other places, *e.g.*, it has thinned by 8 per cent in the area over 30°-50°N latitude between 1979-1990.

The temperature decreases with increasing altitude in the toposphere (8-16 km from earth surface), while it increases with increasing altitude in the stratosphere (above 16km upto 50 km). This rise in temperature in (Stratosphere is caused by the ozone layer. The ozone layer has two important and interrelated effects.

Firstly, it absorbs UV light and thus protects all life on earth from harmful effects of radiations. Second, by absorbing the UV radiation the ozone layer heats the stratosphere, causing temperature inversion. The effect of this temperature inversion is very interesting.

It limits the vertical mixing of pollutants, thereby causing the dispersal of the pollutants over large areas and near the earth's surface. That is why a dense cloud of pollutants usually hangs over the atmosphere in highly industrialised areas causing several unpleasant effects.

Thinning of hole in ozone layer allows harmful UV rays to reach parts of earth. It causes skin cancer, *e.g.*, in Australia over which ozone depletion has been recorded, 70 per cent of the adults reaching the age of 70 suffer from skin cancer. Beside skin cancer there is a supression of immune system, increased susceptibility to herpes, high incidence of cataract and dimming of eye sight. Large scale damage also occurs to green producers both in sea and over land. Food supply will, therefore reduce causing famine. A number of land animals are blinded.

Thinning of ozone layer or hole in ozone layer is caused by a number of pollutants. Major pollutants responsible for depletion are Chlorofluorocarbons (CFCs), nitrogen oxides (coming from fertilizers) carbon tetrachloride, halon and methyl chloroform. Chlorofluorocarbons ($CC1_3F$, $CC1_2 F_2$) have maximum ozone depleting potential or DPD. CFCs are widely used as coolant in air conditioners and refrigerators, cleaning solvents, aerosol propellants and in foam insulation. CFC is also used in fire extinguishing equipment. They escape as aerosol in the stratosphere. In the stratosphere CFCs split up to release chlorine. Chlorine changes ozone to oxygen.

$$CC1_2F_2 \rightarrow CC1F_2 + C1$$

$$C1 + 20_3 \rightarrow C1 + 30_2$$

Chlorine seems to play dominating role in depleting ozone layer. A single chlorine atom is sufficient to convert 1 lakh molecules of ozone into oxygen. Chlorine is being emitted into troposphere and stratosphere by large number of rockets being fired into space. Nitric oxides released into stratosphere by jets also reacts with ozone to form oxygen.

$$NO + O_3 \rightarrow NO_2 + O_2$$

Universal Matter

The first global conference on the depletion of ozone layer was held in Vienna (Austria) in 1985 the year, scientists discovered hole in South Pole. This was followed by Montreal Protocol in 1987 which called for a 50 per cent cut in the use of CFCs by 1998. Many countries including India did not sign the Protocol.

The three day international "Saving the ozone layer" conference was organised jointly in London in March 1989 by the British Govt. and the UNEP which resulted 37 more countries to support for the Montreal Protocol which was initially signed by 31 countries.

In May 1989 another International Conference on Ozone was held at Helsinki. As many as 80 nations agreed to have a total ban on chemicals that cause ozone depletion by 2000 A.D. The agreement for CFC elimination by 2000 A.D. is needed as a major step towards environmental protection.

Pollution being Controlled

Air pollution is the big problem of present society. It is the need of the present day to make the social as well as legislative measures to protect the environment. Three types of steps can be taken to control air pollution. Separation of the pollutants from harmless gases, avoidance of pollutants, and conversion of pollutants to harmless materials.

Separation of the Pollutants: This can be done by the following steps.

(i) Trees should be grown in all available places. The trees use CO_2 and release oxygen. This purifies the air for man and animals to breathe.

(ii) Certain plants such as Ficus variegata, Dancus carrota, phaseolus vulgaris can fix carbon monoxide and some plants like Pinus, Juniperus, Quercus, Pyrus can metabolise nitrogen oxides. Plantation of such species should be encouraged to depollute the air.

(iii) Sulphur free and lead free fuel should be used for motor vehicles. The exhaust gases from motor vehicles may be cleaned by use of catalytic converters.

(iv) The current industrial processes may be suitably modified so as to reduce or check air pollution. Control equipment like gravity settling tanks or porous filters and electrostatic precipitators should be installed in factories to minimise the air pollution.

(v) In factories chimneys should be tall to reduce pollution of air at ground level.

(vi) Industrial smoke should be filtered before releasing it into the air to remove particulate matter.

(vii) Use of generators in residential areas should be avoided.

(viii) Poisonous gases should be removed by passing the fumes through water tower scrubber or spray collector.

(ix) Reformulate gasoline to save ozone in the atmosphere.

(x) Control equipment like Gravity Settling chambers, Porous filters cyclone collectors or electrostatic precipitators should be used in the factories to reduce air pollution.

(a) *Gravity Settling Chambers:* They are long chambers where slow moving gaseous stream deposits its heavier and larger particles (larger than 50 mm).

(b) *Porous or Bag Filters:* They are large sized porous bags of polyester, polyprolene, teflon through which dry exhaust emissions are passed under pressure to filter out particulate matter.

(c) *Cyclone Collectors:* They contain a chamber in which gas stream with particulate matter is whorled round through tight circular spirals. The particulate matter get centrifuged, collected and removed.

(d) *Electrostatic Precipitators (ESPs):* They are devices where air stream with particulate matter is passed through regions having electrically charged plates so that they drift to an electrically grounded wall from where they can be removed easily.

(e) *Wet Scrubbers:* They separate gases by passing air stream through a fine spray of water.

Avoidance of Pollutants: This is done by following measures:

(i) Use of automobiles should be minimised.

(ii) Conventional fuels (Fire wood, coal, oil) should be replaced by electricity or natural gas. These will not emit SO_3

(iii) Population should be brought under control.

(iv) Industries should be away from residential areas.

(v) Pollution free oils should be developed for automobiles.

Conversion of Pollutants: This is done by oxidation in the air or by chemical neutralisation of acid and bases.

Recent steps taken to reduce air pollution are:

(i) The supreme court directed about 168 industries of Delhi, including big industrial houses such as Birla Textile Mills, DCM Silk Mills, Sri Ram Foods and Fertilizers, etc. to shift to those areas by Nov. 1997 which are less hazardous.

(ii) Since Sept., 1997, the Delhi Government has banned plying of those vehicles which are more than 15 years old.

(iii) On Sept. 16, 1997 a Montreal Protocol was signed to identify and stop the use of ozone depleting substances. So Sept. 16 is celebrated by international community as Ozone Day.

(iv) Indian Institute of Petroleum (IIP) has developed new eco-friendly multifuel low air pressure burners in Dec., 1997. These burners would save 10 to 20 per cent of fuel over the conventional burners.

(v) HFC-1340 is a country made ozone-friendly refrigerant chemical deployed by Indian Institute of Chemical Technology (HCT), Hyderabad. It will replace Freon-12 gas which causes depletion of ozone layer.

(vi) Government of India has made it compulsory for all new cars to be fitted with catalytic converters and stated that unleaded petrol will be available throughout the country by 2000 A.D.

Water Pollution: Water Pollution is the degradation of the quality of water due to addition of foreign substances (organic, inorganic, biological or radiological) to water or change in its physical property (temperature) so that it becomes a health hazard and unfit for use.

Water is one of the abundantly available substances in nature. It is an essential constituent of all animals and vegetable matter and forms about 75 per cent of the matter of earth's crust. The earth has about 1.35 cubic kilometers of water of which about 97 per cent is found in the oceans. The main source of land water is atmospheric rain. It is estimated that about 27 per cent of rain water flows into oceans and about 73 per cent is evaporated.

A little percentage of watei enters the soil whose some part reaches the deep zone due to force of gravity and called gravitational water. The upper layer of it called water table. Surface water usually contains small amount of suspended particles (organic and inorganic) and a number of microorganismrs such as bacteria, algae, viruses, protozoas etc. Water becomes polluted when concentration of these increases. Water pollution is a serious health hazard in India. About 50-60 per cent Indian population suffers from water borne diseases and 30-40 per cent all deaths are due to water pollution. Availability of fresh water has declined by 2/3rd in the past 50 years.

Man is the main cause of water pollution. Some pollution occurs naturally too. Soil particles enter water by its erosion;

mineral dissolved in water from rocks and soil; animal wastes and dead fallen leaves fall into water sources, decaying of organic matter also pollutes water.

Division of Pollution

Water sources (Ponds, tanks, lakes, streams and rivers) receive 6 main types of pollutants: organic wastes, pathogenic organisms, inorganic wastes (chemical and minerals), radioactive wastes, solid particles and heat.

Sources: Different sources which add the pollutants in water are broadly classified into following groups

(a) Domestic effluents

(b) Industrial effluents

(c) Surface run off

(d) Thermal pollution (waste heat)

Domestic Effluents: It consists of waste water which is discharged into sewerage system. It contains human and animal excreta, food residues, detergents, organic wastes from tanneries, slaughter houses and canning industries, a large number of bacteria and discharges, from other commercial establishments connected to public sewerage system. Municipal sewers contain many kinds of pollutants and a lot of industrial wastes. The sewage is passed into water courses, generally without treatment.

Beautiful Dal Lake of Kashmir has become polluted as raw sewage and other domestic effluents are passed into it from nearby localities, hotels and house boats. Hussainsagar lake (major source of drinking water for Hyderabad) receives some 30 million litres of waste water daily from domestic and industrial sources. Delhi produces 2000 million litres of sewage, half of which is passed untreated in Yamuna. A time will come when the water of Yamuna will become unfit for human use even after treatment.

A special feature of this is Putrescibility, *.i.e.* decay and decomposition of organic matter present in it due to the presence of bacteria and other micro organisms. Most laundry detergent contain phosphate. People in remote areas often take bath, wash their clothes and animals in the same pond. Such ponds become heavily polluted.

Sewage pollution is due to the presence of coliforms, enterococci, eggs of intestinal worms and pathogens of typhoid, dysentery, diarrhoea, cholera, hepatitis, jaundice, etc. Two-thirds of all illness and one-third of all deaths in India are due to water borne diseases caused by sewage contamination.

IWP are indices of water pollution and are calculated by the sensitivity of Daphnia and Trout to organic wastes in water, while faecal pollution is indicated by number of Escherichia coli in water (unit being mpn).

In water organic matter provides nutrition for decomposers, namely bacteria (Escherichia) and Fungi (Mucor, Leptomitus). They break down the organic matter using oxygen in the process. It results in decreased oxygen content. Water having dissolved oxygen (D.O.) below 8 ppm is polluted. It is heavily polluted below 4 ppm of D.O.

Biological Oxygen Demand: It is the amount of oxygen required in milligrams by decomposer micro organisms in five days to complete the decomposition of organic matter present in one litre of water at 20°C. B.O.D. indicates the quality of waste water, .*i.e.* degree of pollution pure drinking water should have a B.O.D. less than 1ppm. A weak organic waste has B.O.D. below 1500 mg/litre, medium organic pollution has B.O.D. 1500-4000 mg/litre while high organic pollution has B.O.D. above 4000 mg/litre.

Industrial Effluents: They are industrial wastes which are allowed to pass into water bodies. The effluents contain heavy metal such as mercury, lead, copper, arsenic, cadmium, zinc and acids and alkalies. Acids and alkalies destroy micro organisms. Some organic pollutants present in effluent are phenol, naphtha, proteins, cellulose fibres and aromatic compounds.

Some of the chemicals are carcinogens. Industrial effluents are most hazardous pollutants both on land and water. Main source of mercury is combustion of coal, smelting of metallic ores battery and paint industries cadmium is released from mines, metal, welding, electroplating and pesticide industries. Lead in added by smelters, chemical and pesticide industries.

Arsenic contamination of drinking water was reported in West Bengal in the beginning of 1980's Initially only seven villages were reported to be affected but this number has gone upto 840 by the end of 1997. About two million people are using arsenic contaminated water. The main cause is use of underground water through deep tube wells.

Surface Run Off: Run off is flow of extra water present on the surface into water reservoirs and water courses. Since croplands are provided with fertilizers and fields are sprayed with pesticides, the surface run off from these fields brings heavy loads of pollutants into natural water bodies. Pesticides such as DDT are non-degradable. India alone uses about 100,000 tonnes of pesticides annually. Fertilizers include, nitrates, phosphates and sulphates of potassium. They reach the ground water through leaching or carried to rivers, lakes and ponds. The nutrient enriched water show euthrophication.

Thermal Pollution: It is caused by the production of hot effluents, hot air and hot water. Rise in temperature of air and water to a harmful level due to heat from power plant industries and automobiles called thermal pollution. Hot water is released in all those industries which employ water as coolant. Industries employing steam also give out a lot of hot water. Hot water kills both plant and animal life in the area of its discharge, *e.g.*, nuclear power station located in Kota draws cool water from Chambal river and discharges hot water into it.

Effects of Domestic Effluents and Organic Wastes

(1) Water borne pathogens cause a number of diseases like jaundice, typhoid, cholera, dysentery, etc. The usual method to find out their concentration is by calculating

their most probable number (mpn.) According to Central Water Health Engineering Institute, 60 persons out of 100000 die every year due to typhoid, dysentery etc, which are caused by polluted water. Diarrhoea is mostly due to water contamination. It kills 4.6 million children every year.

(2) Sewage stimulates the activity of decomposer micro organisms collectively called sewate fungus. It contains bacteria (*e.g.*, Beggiotoa, Eschrichia), fungi (Mucor, Fusarium) and some algal (chlamy domonas, cladophora, etc.)

(3) Sewage makes the water turbid, brownish, oily and foul smelling and makes it unfit for human consumption.

(4) Organic wastes form a scum and sludge in polluted water which becomes unfit for industrial use.

(5) Detergents present in sewage water contain phosphates some minerals are also released during decomposition of organic matter which stimulates the algal blooming.

Effects of Industrial Wastes

(1) Compounds like mercury, arsenic and lead are neurotoxic and cause number of nervous disorders.

(2) Minimata diseases was first reported in Japan in 1952 which appear due to eating of mercury contaminated fishes from the bay Minamata. The victim develop numbness of lips and limbs, impairment of tactile sense, speech and hearing, narrowing of vision, diarrhoea, haemolysis leading to death.

(3) Mercury inhibits chromosomal disjunction during gamete formation (Ramel, 1974). It therefore, brings about genetic changes.

(4) Cadmium polluted water brings about nausea, vomiting, diarrhoea, cramps, renal damage. It also caused a disease itai-itai (ouch-ouch) in Japan(1947)

(5) Lead contaminated water leads to loss of apetite, anaemia, irritability, damage to liver, kidney and brain.

(6) Excess of nitrates in drinking water cause methaemoglobinaemia. In this nitrite enters the blood, combines with haemoglobin and forms methaemoglobin. It reduces oxygen carrying capacity of blood.

(7) Cobalt contamination causes low blood pressure, bone defects, paralysis, diarrhoea, lung irritation.

(8) Excess use of fluorides leads to fluorosis. It is characterised by mottling of teeth, weak bones, boat-shaped posture and knock knees. A high does of fluoride (100-200 ppm) causes the retarded growth.

Effect of Surface Runoff

(1) Excess of nitrates and phosphates from fertilizers are washed into water bodies. It leads to eutrophication.

Eutrophication is the phenomenon of nutrient enrichment of a water body that initially supports a dense growth of plant and animal life. The rapid increased growth of water plants especially algal is called bloom. Bloom generally occurs on the surface. They cut of light from submerged plants. As a result oxygen replenishment decreases inside the water. Night time respiration of animals and plants further decrease the dissolved oxygen. It causes death of a number of animal and submerged plants. Number of decomposers increases which release pollutants like methane, hydrogen sulphide and ammonia which combine to form scum and sludge killing the bloom forming plants. The water body gives foul smell, brown colouration, bad taste and becomes unfit for drinking.

(2) Excess of pesticides, weedicides into water bodies cause Bio-magnification. Bio-magnification or biological concentration or biological amplification in the phenomena of increase in concentration of persistent pesticides per unit weight of organisms with the rise in trophic level of a food chain.

Pesticides are the chemicals used to kill plant and animal pests. These include insecticides, fungicides algicides, rodenticides and weedicides or herbitides. Most of them are broad spectrum and effect, all types of organisms so collectively called biocides. Most of the biocides are non-biodegradable and toxicants. The long range effects of such biocide are infect a threat to our ecological security. According to person (1985) pesticides related death in developing countries are estimated at 10,000 year and about 1.5-2 million persons suffer from acute pesticide poisoning.

Some of most toxic biocides are DDT (Dichlorodipheny trichloroethane), BHC (benzene hexachloride), Chlordane, heptachlor, methoxychlor, toxaphene, aldrin, endrin and PCPs (poly chlorinated biphenyls). Indiscriminate use of the biocides could make them an integral part of our biological geological and chemical cycle of the earth, *e.g.*, if DDT enters a pond, lake, it is taken as such by the plants of the pond, then reaches in 200 plankton feeding on plants, then to fish and finally in the body of the bird who eat the fish.

DDT concentration continuously increases in successive trophic levels in a food chain. This is bio-magnification or biological amplification. This is the reason why our food grains as wheat and rice and vegetables and fruits today contain varying amount of pesticides which have become their integral part and cannot be removed by washing or other means.

Effects of Bio-magnification

(1) In India, endemic familial arthritis (pain in joints, hips and inability to stand up) appeared in Malnad region of Karnataka due to eating of crabs picked from rice fields sprayed with pesticides.

(2) DDT interferes the egg shell formation in many birds. The shell remain thin and break by bird's weight during incubation.

(3) Persistent pesticides are often teratogens and carcinogens besides being poisonous. In higher

concentration they cause softening of brain, cerebral haemorrhage, hypertension and malfunctioning of sex hormones.

(4) Dieldrin is 5 times more toxic than DDT when ingested and 40 times more poisonous when absorbed. Endrin is the most toxic amongst chlorinated hydrocarbons. Endrin rich agricultural wastes killed 5 million fish in one case.

(5) Excessive spray of hard biocides, sometimes causes an imbalance in prey predator population.

Effect of Thermal Pollution

(1) Warmer water contains less oxygen (14 ppm of 0°C and 1 ppm at 20°C) so thermal pollution causes deoxygenation of water therefore, the rate of decay of organic matter falls down and kills the aquatic animal.

(2) High temperature also denatures enzymes, increases respiration but lowers down the rate of photosynthesis. Hence there is decreased primary production.

(3) Troust eggs fail to hatch and Salman does not spawn at higher temperature.

Pollution of water can be checked or at least minimised by the following measures:

(1) Taking bath and washing clothes directly in ponds, tanks and other sources which supply drinking water should be prohibited.

(2) Separate ponds should be reserved for the water supply to cattle and other animals.

(3) Over use of fertilizer and pesticides should be avoided.

(4) Solid wastes should be recycled where ever possible.

(5) Hot water should be cooled before release from factories.

(6) Reverse osmosis in which brackish water is demineralised by pumping it through a semi-permeable membrane under strong pressure.

(7) Water hyacith can purify water by taking up heavy metals, such as lead, mercury, cadmium and nickel and some toxic material from water.

(8) Domestic and farm yard sewage and industrial waste should be suitably treated before rinsing them in water. This process can reduce the harmful effect of the wastes. It is given below:

Waste Water Treatment: Waste water is the water which has been used for a purpose and cannot be employed again unless and until it has been removed of pollutants or impurities that have crept in during the previous use. Waste water carry a number of pollutants.

The two common sources of water pollution are sewage and industrial effluents. All water courses of India are badly polluted due to passage of waste water into them from human settlements carrying untreated sewage and industrial effluent.

Central Ganga Authority was established in 1985 to free Ganga from Sewage and effluent contamination. It is planned to install sewage treatment plants for 27 cities on the bank of Ganga to handle about 1,000 million litres of sewage daily before its discharge into river.

Ganga Action plant (GAP) was launched in June 14,1986. Similar action plan has been started for cleaning the Yamuna and Gomti rivers in which river Yamuna covers 15 big cities and Gomti covers 3 cities.

Sewage Treatment: The sewage is taken to sewage treatment sites. The treatment occurs in three steps primary, secondary and tertiary.

Primary Treatment: It is also called physical treatment because in this method mechanical screening and sedimentation of undisolved solids in raw sewage (*e.g.* large lumps of organic matter, sand and silt) is done. It fails to remove any dissolved substances in water. It does not remove pathogens.

Secondary Treatment: It is mainly biological treatment because the organic matter is decomposed with the help of microbes the waste water is then sterilised through chlorination. Secondary treatment consists of two steps, decomposition and chlorination.

(i) *Decomposition of Organic Wastes:* There are two methods for decomposing organic wastes.

(a) *Trickling Filter Method:* In this case, sewage after primary treatment is passed through a thick layer of gravel (small stones). Bacteria consume the organic matter during its filtration. The water that trickles out through the bottom of the gravel bed is much cleaner.

(b) *Activated Sludge Method:* The waste water is pumped into an aeration tank. Here it is mixed with air and sludge (consisting of bacteria and algae). The bacteria decompose most of the organic matter while algae provides oxygen for bacteria. The water which is now almost clear of organic matter still carries large amount of nitrates, phosphates, etc.

(ii) *Chlorination:* After decomposition the water is sent by pipes into chambers where it is chlorinated. It kills micro organisms of sewage fungus as well as pathogen contaminants. The water is rich in phosphorus, nitrogen and minerals. It is directly supplied to the field as mannured water.

Tertiary Treatment: It is undertaken when water is to be recycled. In this, nitrates and phosphates are removed by precipitation, filternation, aeration and purification methods. The treated water now can been used or released into natural waters.

Treatment of Industrial Effluents: It consists of neutralisation of acids and alkalies, removal of toxic compounds by chemical oxidation coagulation of colloidal impurities, precipitation of metallic compounds and cooling of waste water.

Soil Pollution: Soil supports plant life, which, in turn, supports animal life. Hence soil pollution affects all organisms. Soil pollution can be defined as alteration in soil by addition and removal of materials leading to decrease in soil fertility. Substances which reduce productivity of the soil are regarded soil pollutants.

Many materials adversely affect the physical, chemical and biological properties of the soil and reduce its productivity. These include:

Domestic Wastes: These are discarded materials or unless leftovers. They include waste foods, paper, clothes, leather, bottles, cans, plastics, ash, etc.

Industrial Wastes: Both solid and liquid wastes of the industry are dumped over the soil. The waste contain scraps, flyash, dyes, plastics, toxic chemical like mercury, copper, zinc, lead, cadmium, cyanides, alkalies, organic solvents, etc. These come from industries involved in manufacture of paper, chemicals, rubber, petroleum, products, cement, sugar, refineries, etc.

Fertilizers and Manure: Fertilizers and manure are added to the soil for increasing crop yield. Excessive use of chemical fertilizers reduces soil's productivity by decreasing its bacterial population and increases its salt content.

Pesticides: These are sprayed on the crops to protect them from pests. These include insecticides (BHC, DDT, aldrin endrin), fungicides and herbicides some of the sprayed pesticides fall on the soil and penetrate inside.

Dumpling of Human Excreta: and wastes from cow sheds and slaughter houses befouls the land. Human excreta also contains many pathogens which cause many soil borne diseases in plants and animals including man.

Ash: It is a solid or powdery mass left after burning of wood, dung and coal. It contains minerals, some of them in toxic concentrations. The ash, therefore, not only renders the nearby soil unfit for agriculture but also pass a number of heavy metals and other toxic chemicals in water and food chain.

Soil pollution is of two main types.

(i) Positive and

(ii) Negative.

Positive Soil Pollution: Reduction in productivity of the soil due to addition of some unwanted materials (industrial wastes, pesticides, discarded materials, etc.) in the soil is called positive, soil pollution. It caused by:

(1) *Salination of Soil:* Increase in the concentration of soluble salts in the soil is called salination. It is thus, a positive soil pollution. It result from.

(i) *Poor Drainage:* The salts dissolved in irrigation and flood water accumulate on soil surface due to poor drainage.

(ii) *Capillary Action:* The salt present in deeper strata are drawn up by capillary action and deposited on the surface.

(iii) *Saline Irrigation Water:* The ground in arid regions is often saline. If used for irrigation, it adds salt to the soil.

(iv) *Parent Rock:* The soil formed by weathering of saline rocks is found to be saline.

(2) Addition of pesticides (DDT, BHC, aldrin, endrin, malathion, pyrethrum) and weedicides in the soil.

(3) *Industrial Wastes:* These are dumped over the soil.

(4) *Mine Dust:* This is major source of pollution in mining areas.

Negative Soil Pollution: It is the loss of soil productivity by reduction in its some useful components or by destruction of its top layer.

(1) *Soil Erosion:* It is the removal of top fertile soil by

agencies like wind, water, etc. converting the latter into desert.

(2) *Reduction in Mineral Contents:* The mineral contents are reduced due to intensive agriculture, flowing water, faulty irrigation and overgrazing.

Effects of Soil Pollution

(1) The chemical and pesticides after the basic composition of the soil. This may kill the essential soil organisms which contribute to structure and fertility of the soil.

(2) The industrial pollutants increase the toxicity level of the soil which proved fatal. In Japan people died of disease itai-itai due to cadmium poisoning of the soil.

(3) The use of inorganic fertilizers spoils the quality of the soil in the long run. They cause accumulation of nitrates in the soil which may cause cyanosis or blue baby syndrome.

(4) Mine dust causes many types of deformities in animals and human beings.

(5) The use of human and animal excreta as manure pollutes the soil besides promoting crop yield.

(6) Excreta may contain pathogens that contaminate the soil and vegetable crops and affect the health of man and domestic animals.

(7) Radioactive dust may find its way from the soil into crops, live stock and humans via food chains.

(8) Excess of chemical fertilizers reduces natural bacterial population in the soil.

(9) Nitrates of chemical fertilizers cause methaemoglobinaemia in man.

(10) Soil pollution also causes a number of plant diseases.

Control of Soil Pollution: Control of soil pollution mainly involves the disposal of solid wastes to provide some benefit to society.

The various methods are:

(1) Solid wastes should be recycled. It is treatment of waste to regenerate a resource. Recycling of newsprint and other waste paper helps reduce pressure on forests. Similarly recycled glass, plastics, polythene, metals, etc., save a lot of energy as well as original resources.

(2) An efficient system of disposal should be developed to deal with domestic solid wastes.

(3) Use of chemical fertilizers and pesticides should be highly judicious.

(4) The use of bio fertilizer and manures can decrease the need for chemical fertilizers which will reduce soil pollution.

(5) Proper legislation should be passed and strictly enforced. Stringent laws should be imposed on defaulters.

(6) Improvement in mining techniques and transport of extracted materially can reduce the spread of mine dust.

(7) Reforestation and plantation can check soil erosion and the advancement of deserts.

(8) In January 1998, Indian scientists have developed a novel method of decontaminating effluents from pulp and paper industry (*e.g.*, chemical oxygen) using chemical wastes (like hyposludge, alum-sludge and bamboo dust carbon) from the same factory.

(9) Proper cropping pattern can eliminate weeds and this can éxclude the need for herbicide use.

Noise Pollution: Sound is the main means of communication in many animals including humans. A low sound is pleasant and harmless. Noise pollution is the release of unwanted, irritating and often excessively higher level of sound. Aloud unpleasant sound is called noise which produces unpleasant effects on the ears.

Frequency of sound is measured in Hertz or Hz. A frequency of 1Hz means one cycle per second. Human beings have a hearing range of 20Hz to 20000 Hz. The unit of loudness of sound intensity is called decibel or dB with zero as the limit of hearing. Moderate conversation has a noise value 60 dB. Sound becomes polluting noise at about 80 dB and becomes intolerable above 100 dB. Noise pollution originates from a number of sources such as:

(1) Domestic gadgets like mixer, pressure cookers, washing machines, desert coolers, air conditions, fans, vacuum cleaners generators (a high as 100dB),etc.

(2) Loud speakers (a noise 60-80 dB in morning 80-100 dB in evening.)

(3) Personal entertainment sources like transistor, radio, record player, T.V, etc.

(4) Agricultural equipment such as harvesters, threshers, tractors, pump sets, etc.

(5) Utensils.

(6) Transport vehicles like scooters, motor cycles, cars, buses, trucks, trains, aeroplanes, helicopters. Air ports produce the maximum noise during landing and take off of aeroplanes. Road transport is major of noise pollution in cities and towns.

(7) Dynamiting of mountains, bulldozers, road roller.

(8) Defence equipment like aircraft, rocket launching, tanks, artillery, etc.

(9) Commercial establishments such as music T.V., exhaust fans, air conditioners, coolers, etc.

(10) Crackers.

Noise pollution affects the power of hearing as well as general health of man. The various affects of noise pollution are:

(1) Noise affects hearing ability. A continuous exposure of noise of 80 dB impairs hearing by 15 dB. The city noise

is generally more than this loudness, therefore, city dwellers are prone to deafness with advancing age. A sudden loud noise may result in rupturing of ear drums.

(2) Noise pollution may cause increased rate of heart beat, blood pressure by increasing level of cholesterol in the blood.

(3) Persons affected by noise pollution suffer from gastric spasms, nausea, pectic ulcers.

(4) Damage to heart, brain and liver has been reported in animals due to prolonged noise pollution.

(5) Constant high level noise results in insomnia (sleeplessness) and headache.

(6) Noise also affects the developing embryo mother's uterus and impair the development CNS of unborn babies.

(7) Noise pollution causes dilation of eye pupil, defective eye sight and colour vision.

(8) Noise also detracts attention and causes emotional disturbances.

(9) Noise pollution causes dilation of blood vessels of brain results in headache.

(10) Noise pollution interferes with our conversation, disturbs concentration and upsets our mood.

Control of Noise Pollution: Three types of measures can be adopted to control noise pollution:

(a) Reduction of noise at sources

(b) Interruption of path of transmission of noise and

(c) Protecting the receiver. The following measures can control noise pollution.

 (1) Silence zones should be created around educational institutions, hospitals and residential areas.

 (2) Noise producing machinery or their parts are covered by sound insulating materially or kept in side sound proof-rooms.

(3) Noise producing industries, railway stations, aerodromes should be located away human settlements.

(4) Radios and transistors should be kept at low volume.

(5) All noise producing machines should be replaced by quieter machines.

(6) Proper laws should be enforced to check the misuse of loudspeakers and announcement systems.

(7) Use of pressure horns should not be allowed in towns and cities.

(8) The intense sound produced by jets and other objects should be deflected away from residential areas.

(9) Motor vehicle noise on roads can be reduced by planting many rows of trees.

(10) Proper lubrication, repair and maintenance of machinery reduce the noise.

Deterioration of Ozone Layer

The ozone is a clear, blue gas with a sharp smell. The ozone layer in the upper atmosphere (stratosphere) protects the living organisms from ultraviolet(UV) rays of the sun by absorbing nearly all of them. The ozone formed in the lower atmosphere by photochemical reaction as a result of human activity is harmful. At low concentration, it produces chest pain, coughing and often eye irritation.

High concentration can kill both plants and animals. Ozone toxicity of plant is manifested by marking on leaves and leads to premature yellowing and fall of leaves. It leads to discoloration and damage to textiles at 1 ppm. Ozone injures mucous membrane. 16th September, 1996 was celebrated as "International Day for the Prevention of Ozone Layer." It was aimed to create awareness about the danger of ozone depletion in the stratosphere.

Peroxyacyl Nitrates (PAN): It causes irritation of eyes and throat, and produces respiratory troubles (asthma, bronchitis, lung cancer) in man. It affects the plant, particularly leafy vegetables such as spinach and lettuce.

Aldehyde: This produces irritation in gastrointestinal and respiratory tracts.

Phenols: They cause damage to liver, lungs, etc.

Smog: Smog (Smoky fog) is a thick, dark or yellow fog over a town produced by combination of smoke and fog. It causes glazing, silvering and necrosis of crops. It also produces respiratory problems in humans and reduces visibility leading to accidents.

Smog is of two types—Chemical smog (contains particulate matter and SO_2) and photochemical smog (contains pollutants like O_3, oxides of nitrogen, CO and hydrocarbons.

Benzpyrene: It is produced in tobacco smoke. Benzpyrene also results from coal and oil combustion. It occurs in automobile exhaust too. It causes lung cancer.

Ethylene: It is an unsaturated hydrocarbon produced from incomplete combustion of coal, wood, petrol, cigars, etc. It causes premature senescence and abscission.

Indoor Pollution: It is caused by combustion of fuels and tobacco smoking. It is responsible for increased respiratory disorders and poor lung functioning.

Particulate Matter: They are added to air by industries and automobiles and by operations such as blasting, drilling, crushing, grinding, mixing, etc. Some particulate matter is added to the air by living organisms. It comprises pollen, spores, cysts and bacteria.

Particulate matter is of two types: Settable (large particles such as sand and water droplets which quickly settle down in still air) and suspended (very fine particles such as tobacco smoke do not settle at all). They are known as non-settleable particles.

The later again classified into 3 categories:

Aerosols—less than 1 mm in diameter are solid or liquid.

Dust—Over 1 mm in diameter and are solid.

Mist—liquid more than Immune diameter.

10-15 per cent of the air pollution is caused by particulate matter.

Effects on Livingbeings: The effect of particulate pollutants depends on the size of the particles. Particles larger than 2 mm are trapped in nasal hair and bronchial mucus, whence they are passed out.

Smaller particles reach the lung alveoli with inhaled air. Here they may be engulfed by phagocytes or absorbed into the blood and prove harmful.

(i) Air borne organic materials, such as spores, pollen, bacteria, fungi, fur, feathers produce allergic reactions, bronchial asthma, tuberculosis and lung cancer.

(ii) Dust, smoke and smog cause bronchitis, asthma and lung diseases. Cotton dust causes lung fibrosis. Asbestos fibres cause lung cancer.

(iii) The effect of gaseous pollutants depends upon their solubility in water, which allows their diffusion into the tissue.

(a) Sulphur dioxide causes drying of the mouth, sore throat and eye irritation. It may damage the tissue by forming sulphuric acid.

(b) Sulphur trioxide, nitrogen oxide and carbon monoxide combine with haemoglobin of the blood and reduce its oxygen carrying capacity leading to hypoxia in body tissues. It causes headache, muscular weakness, nausea, exhaustion. Nitrogen oxide impairs the working of lungs by causing accumulation of water in the air spaces.

NO, inhalation causes eye irritation, lung oedema, blood congestion.

(c) Hydrocarbons are known to cause eye-irritation, coughing, sneezing, drowsiness, etc. Benzene is a carcinogen causing leukaemia.

(d) Lead affects central nervous system and distorts red blood corpuscles.

(e) SPM penetrates deep into respiration system and cause bronchitis, asthma, cardiovascular problems.

(f) Benzpyrena of tobacco smoke is arcinogenic.

(g) Peroxyacyl nitrate causes irritation of eyes and throat and respiratory diseases.

(h) Phenols damage spleen, kidneys, liver and lungs.

(i) Chlorofluorocarbons (CFCs) undergo bio-magnification and may reach upto a level of 1400mg/kg of pectoral muscles in sea eagle and 3.5 mg/kg of fat in mother's milk.

These impair reproductive activities of animals and damage the liver, CNS, vision, etc. in human beings.

(j) PCBs damage liver and CNS impair vision and change skin pigmentation.

(k) Ozone damages the mucous membrane at a concentration of even less than 1 ppm.

(l) Aldehydes cause irritation in gastrointestinal and respiratory tracts.

(m) Smog causes respiratory disorder like asthma, allergies, etc.

Effects on Plants

(i) SO_2 causes chlorosis, plasmolysis, membrane damage, metabolic inhibition and death. It does a great harm to forest tree.

(ii) Fluorides and PAN damage leafy vegetables such as spinach and lettuce.

(iii) Ozone and hydrocarbons cause premature yellowing and fall of leaves and flower buds and discoloration and curling of sepals.

(iv) Nitrogen oxides reduce yield of crops.

(v) Dust, Smoke and Smog reduce sunlight and form a thin layer on the leaves thereby retarding photosynthesis.

(vi) Lichens are sensitive to air pollution. They are grown as pollution indicators.

2

WEALTH FROM EARTH

Land as a Resource: Land is a finite and valuable resource upon which we depend for our food, fibre and fuel wood, the basic amenities of life. Soil, especially the top soil, is classified as a renewable resource because it is continuously regenerated by natural process though at a very slow rate. About 200-1000 years are needed for the formation of one inch or 2.5 cm soil, depending upon the climate and the soil type. But, when rate of erosion is faster than rate of renewal, then the soil becomes a non-renewable resource.

Land Degradation: With increasing population growth the demands for arable land for producing food, fibre and fuel wood is also increasing. Hence there is more and more pressure on the limited land resources which are getting degraded due to over-exploitation. Soil degradation is a real cause of alarm because soil formation is an extremely slow process as discussed above and the average annual erosion rate is 20-100 times more than the renewal rate. Soil erosion, water-logging, salinisation and contamination of the soil with industrial wastes like fly-ash, press-mud or heavy metals all cause degradation of land.

Soil Erosion: The literal meaning of 'soil erosion' is wearing away of soil. Soil erosion is defined as the movement of soil components, especially surface-litter and top soil from one place to another.

Soil erosion results in the loss of fertility because it is the top soil layer which is fertile. If we look at the world situation, we find that one-third of the world's cropland is getting eroded. Two-third of the seriously degraded lands lie in Asia and Africa.

Soil erosion is basically of two types based upon the cause of erosion:

(i) *Normal Erosion or Geologic Erosion:* caused by the gradual removal of top soil by natural processes which bring an equilibrium between physical, biological and hydrological activities and maintain a natural balance between erosion and renewal.

(ii) *Erosion at Accelerated Pace:* This is mainly caused by anthropogenic (man-made) activities and the rate of erosion is much faster than the rate of formation of soil. Overgrazing, deforestation and mining are some important activities causing accelerated erosion. There are two types of agents which cause soil erosion:

    (a) *Climatic Agents:* water and wind are the climatic agents of soil erosion. Water affects soil erosion in the form of torrential rains, rapid flow of water along slopes, runoff, wave action and melting and movement of snow.

Water induced soil erosion is of the following types:

- *Sheet Erosion:* when there is uniform removal of a thin layer of soil from a large surface area, it is called sheet erosion. This is usually due to runoff water.
- *Rill Erosion:* When there is rainfall and rapidly running water produces finger-shaped grooves or rills over the area, it is called rill erosion.

- *Gully Erosion:* It is a more prominent type of soil erosion. When the rainfall is very heavy, deeper cavities or gullies are formed, which may be U or V shaped.
- *Slip Erosion:* This occurs due to heavy rainfall on slopes of hills and mountains.
- *Stream Bank Erosion:* During the rainy season, when fast running streams take a turn in some other direction, they cut the soil and make caves in the banks.

Wind erosion is responsible for the following three types of soil movements:

- *Saltation:* This occurs under the influence of direct pressure of stormy wind and the soil particles of 1-1.5 mm diameter move up in vertical direction.
- *Suspension:* Here fine soil particles (less than 1 mm diameter) which are suspended in the air are kicked up and taken away to distant places.
- *Surface Creep:* Here larger particles (5-10 mm diameter) creep over the soil surface along with wind.

(b) *Biotic Agents:* Excessive grazing, mining and deforestation are the major biotic agents responsible for soil erosion. Due to these processes the top soil is disturbed or rendered devoid of vegetation cover. So the land is directly exposed to the action of various physical forces facilitating erosion. Overgrazing accounts for 35% of the world's soil erosion while deforestation is responsible for 30% of the earth's seriously eroded lands. Unsustainable methods of farming cause 28% of soil erosion.

Deforestation without reforestation, overgrazing by cattle, surface mining without land reclamation, irrigation techniques that lead to salt build-up, water-logged soil, farming on land with unsuitable terrain, soil compaction by agricultural machinery, action of cattle trampling, etc. make the top soil vulnerable to erosion.

Protection of Earth

- While constructing your house, don't uproot the trees as far as possible. Plant the disturbed areas with a fast-growing native ground cover.
- Grow different types of ornamental plants, herbs and trees in your garden. Grow grass in the open areas which will bind the soil and prevent its erosion.
- Make compost from your kitchen waste and use it for your kitchen-garden or flower-pots.
- Do not irrigate the plants using a strong flow of water, as it would wash off the soil.
- Better use sprinkling irrigation.
- Use green manure and mulch in the garden and kitchen-garden which will protect the soil.
- If you own agricultural fields, do not over-irrigate your fields without proper drainage to prevent water-logging and salinisation.
- Use mixed cropping so that some specific soil nutrients do not get depleted.

Durable Agriculture Incentives

- Do not waste food. Take as much as you can eat.
- Reduce the use of pesticides.
- Fertilize your crop primarily with organic fertilizers.
- Use drip irrigation to water the crops.
- Eat local and seasonal vegetables. This saves lot of energy on transport, storage and preservation.
- Control pests by a combination of cultivation and biological control methods.

Resources use Equitable Term for Sustainable Life-style: There is a big divide in the world as North and South, the more developed countries (MDC's) and less developed countries (LDC's), the haves and the have-nots. The less developed does not mean that they are backward as such, they are culturally very rich or even much more developed, but economically they are less developed. The gap between the two is mainly because of population and resources.

The MDC's have only 22% of world's population, but they use 88% of its natural resources, 73% of its energy and command 85% of its income. In turn, they contribute a very big proportion to its pollution. These countries include USA, Canada, Japan, the CIS, Australia, New Zealand and Western European Countries. The LDC's, on the other hand, have very low or moderate industrial growth, have 78% of the world's population and use about 12% of natural resources and 27% of energy. Their income is merely 15% of global income. The gap between the two is increasing with time due to sharp increase in population in the LDC's. The rich have grown richer while the poor have stayed poor or gone even poorer.

As the rich nations are developing more, they are also leading to more pollution and sustainability of the earth's life support system is under threat. The poor nations, on the other hand, are still struggling hard with their large population and poverty problems. Their share of resources is too little leading to unsustainability.

As the rich nations continue to grow, they will reach a limit. If they have a growth rate of 10 % every year, they will show 1024 times increase in the next 70 years. Will this much of growth be sustainable? The answer is 'No' because many of our earth's resources are limited and even the renewable resources will become unsustainable if their use exceeds their regeneration.

Thus, the solution to this problem is to have more equitable distribution of resources and wealth. We cannot expect the poor countries to stop growth in order to check pollution because development brings employment and the main problem of these countries is to tackle poverty. A global consensus has to be reached for more balanced distribution of the basic resources like safe drinking water, food, fuel, etc. so that the poor in the LDC's are at least able to sustain their life. Unless they are provided with such basic resources, we cannot think of rooting out the problems related to dirty, unhygienic, polluted, disease-infested settlements of these people which contribute to unsustainability.

Thus, the two basic causes of unsustainability are over population in poor countries who have under-consumption of resources and over-consumption of resources by the rich countries, which generate wastes.

In order to achieve sustainable life styles it is desirable to achieve a more balanced and equitable distribution of global resources and income to meet everyone's basic needs.

The rich countries will have to lower down their consumption levels while the bare minimum needs of the poor have to be fulfilled by providing them resources. A fairer sharing of resources will narrow down the gap between the rich and the poor and will lead to sustainable development for all and not just for a privileged group.

Energy Preservation

- Turn off lights, fans and other appliances when not in use.
- Obtain as much heat as possible from natural sources. Dry the clothes in sun instead of drier if it is a sunny day.
- Use solar cooker for cooking your food on sunny days which will be more nutritious and will cut down on your LPG expenses.
- Build your house with provision for sunspace which will keep your house warmer and will provide more light.
- Grow deciduous trees and climbers at proper places outside your home to cut off intense heat of summers and get a cool breeze and shade. This will cut off your

electricity charges on coolers and air-conditioners. A big tree is estimated to have a cooling effect equivalent to five air conditioners. The deciduous trees shed their leaves in winter. Therefore they do not put any hindrance to the sunlight and heat.

- Drive less, make fewer trips and use public transportations whenever possible. You can share by joining a car-pool if you regularly have to go to the same place.
- Add more insulation to your house. During winter close the windows at night. During summer close the windows during days if using an A.C. Otherwise loss of heat would be more, consuming more electricity.
- Instead of using the heat convector more often wear adequate woolens.
- Recycle and reuse glass, metals and paper.
- Try riding bicycle or just walk down small distances instead of using your car or scooter.
- Lower the cooling load on an air-conditioner by increasing the thermostat setting as 3-5 % electricity is saved for every one degree rise in temperature setting.

Soil Preservation

In order to prevent soil erosion and conserve the soil the following conservation practices are employed:

(i) *Conservational till Farming:* In traditional method the land is ploughed and the soil is broken up and smoothed to make a planting surface. However, this disturbs the soil and makes it susceptible to erosion when fallow (i.e. without crop cover). Conservational till farming, popularly known as no-till-farming causes minimum disturbance to the top soil. Here special tillers break up and loosen the subsurface soil without turning over the topsoil. The tilling machines make slits in the unploughed soil and inject seeds, fertilizers, herbicides and a little water in the slit, so that the seed germinates and the crop grows successfully without competition with weeds.

(ii) *Contour Farming:* On gentle siopes, crops are grown in rows across, rather than up and down, a practice known as contour farming. Each row planted horizontally along the slope of the land acts as a small dam to help hold soil and slow down loss of soil through runoff water.

(iii) *Terracing:* It is used on still steeper slopes are converted into a series of broad terraces which run across the contour. Terracing retains water for crops at all levels and cuts down soil erosion by controlling run off. In high rainfall areas, ditches are also provided behind the terrace to permit adequate drainage.

(iv) *Strip Cropping:* Here strips of crops are alternated with strips of soil saving covercrops like grasses or grass-legume mixture. Whatever runoff comes from the cropped soil is retained by the strip of cover-crop and this reduces soil erosion. Nitrogen fixing legumes also help in restoring soil fertility.

(v) *Alley Cropping:* It is a form of inter-cropping in which crops are planted between rows of trees or shrubs. This is also called agro-forestry. Even when the crop is harvested, the soil is not fallow because trees and shrubs still remain on the soil holding the soil particles and prevent soil erosion.

Wind breaks or shelterbelts: They help in reducing erosion caused by strong winds. The trees are planted in long rows along the cultivated land boundary so that wind is blocked. The wind speed is substantially reduced which helps in preventing wind erosion of soil.

Thus, soil erosion is one of the world's most critical problems and, if not slowed, will seriously reduce agricultural and forestry production, and degrade the quality of aquatic ecosystems as well due to increased siltation. Soil erosion, is in fact, a gradual process and very often the cumulative effects becomes visible only when the damage has already become irreversible. The best way to control soil erosion is to maintain adequate vegetational cover over the soil.

Water-logging: In order to provide congenial moisture to the growing crops, farmers usually apply heavy irrigation to their farmland. Also, in order to leach down the salts deeper into the soil, the farmer provides more irrigation water. However, due to inadequate drainage and poor quality irrigation water there is accumulation of water underground and gradually it forms a continuous column with the water table. We call these soils as waterlogged soils which affect crop growth due to inhibition of exchange of gases. The pore-spaces between the soil particles get fully drenched with water through the roots.

Water-logging is most often associated with salinity because the water used for irrigation contains salts and the soils get badly degraded due to erroneous irrigation practices. The damages caused by some major irrigation projects is shown is below:

Water-logging and Salinisation caused due to Some Irrigation Projects in India

Irrigation Project	*State*	*Area affected (thousand hectares)*	
		Water logging	*Salinity*
Indira Gandhi Canal	Rajasthan	43	29
Gandak	Bihar	211	400
Chambal	M.P., Rajasthan	98	40
Ram Ganga	UP	195	352
Sri Ram Sagar	Andhra Pradesh	60	1

Source: B.K. Garg and I.C. Gupta (1997).

An estimated loss of Rs. 10,000 million per annum occurs due to water-logging and salinity in India. It is a startling fact because the cost of development of the irrigation projects is very high and in the long run they cause problems like water-logging and salinity thereby sharply reducing soil fertility.

Landslides: Various anthropogenic activities like hydroelectric projects, large dams, reservoirs, construction of roads and railway lines, construction of buildings, mining, etc. are responsible for clearing of large forested areas. Earlier there were few reports of landslides between Rishikesh and Byasi on Badrinath Highway area. But, after the highway was constructed, 15 landslides occurred in a single year.

During construction of roads, mining activities, etc. huge portions of fragile mountainous areas are cut or destroyed by dynamite and thrown into adjacent valleys and streams. These land masses weaken the already fragile mountain slopes and lead to landslides. They also increase the turbidity of various nearby streams, thereby reducing their productivity.

Desertification: Desertification is a process whereby the productive potential of arid or semiarid lands falls by ten per cent or more. Moderate desertification is 10-25% drop in productivity, severe desertification causes 25-50% drop while very severe desertification results in more than 50% drop in productivity and usually creates huge gullies and sand dunes. Desertification leads to the conversion of rangelands or irrigated croplands to desert-like conditions in which agricultural productivity falls.

Desertification is characterised by de-vegetation and loss of vegetable cover, depletion of groundwater, salinisation and severe soil erosion. Desertification is not the literal invasion of desert into a non-desert area. It includes degradation of the ecosystems within as well as outside the natural deserts. The Sonoran and Chihuahuan deserts are about a million years old, yet they have become more barren during the last 100 years. So, further desertification has taken place within the desert.

Causes of Desertification: Formation of deserts may take place due to natural phenomena like climate change or may be due to abusive use of land. Even the climate change is linked in many ways to human activities. The major anthropogenic activities responsible for desertification are as follows:

(a) *Deforestation:* The process of denuding and degrading a forested land initiates a desert producing cycle that feeds on itself. Since there is no vegetation to hold back the surface runoff, water drains off quickly before it can soak into the soil to nourish the plants or to replenish the groundwater. This increases soil erosion, loss of fertility and loss of water.

(b) *Overgrazing:* The regions most seriously affected by desertification are the cattle-producing areas of the world. This is because the increasing cattle population heavily graze in grasslands or forests and as a result denude the land area. When the earth is denuded, the microclimate near the ground becomes inhospitable to seed germination. The dry barren land becomes loose and more prone to soil erosion. The top fertile layer is also lost and thus plant growth is badly hampered in such soils. The dry barren land reflects more of the sun's heat, changing wind patterns, driving away moisture-laden clouds leading to further desertification.

(c) *Mining and Quarrying:* These activities are also responsible for loss of vegetal cover and denudation of extensive land areas leading to desertification. Deserts are found to occur in the arid and semi-arid areas of all the Continents. During the last 50 years about 900 million hectares of land have undergone desertification over the world. This problem is especially severe in Sahel region, just south of the Sahara in Africa. It is further estimated that if desertification continues at the present rate, then by 2010, it will affect such lands which are presently occupied by 20% of the human population.

Amongst the most badly affected areas are the sub Saharan Africa, the Middle East, Western Asia, parts of Central and South America, Australia and the Western half of the United States. It is estimated that in the last 50 years, human activities have been responsible for desertification of land area equal to the size of Brazil. The UNEP estimates suggest that if we don't make sincere efforts now then very soon 63% of rangelands, 60% of rain-fed croplands and 30% of irrigated croplands will suffer from desertification on a worldwide scale, adding 60,000 Km^2 of deserts every year.

Natural Resources' Conservation Individual's Part: Different natural resources like forests, water, soil, food, mineral and energy resources play a vital role in the development of a nation. However, overuse of these resources in our modern society is resulting in fast depletion of these resources and several related problems. If we want our mankind to flourish there is a strong need to conserve these natural resources.

While conservation efforts are underway at national as well as international level, the individual efforts for conservation of natural resources can go a long way. Environment belongs to each one of us and all of us have a responsibility to contribute towards its conservation and protection. "Small droplets of water together form a big ocean". Similarly, with our small individual efforts we can together help in conserving our natural resources to a large extent. Let us see how can individuals help in conservation of different resources.

Water Preservation

- Don't keep water taps running while brushing, shaving, washing or bathing.
- In washing machines fill the machine only to the level required for your clothes.
- Install water-saving toilets that use not more than 6 liters per flush.
- Check for water leaks in pipes and toilets and repair

them promptly. A small pin-hole-sized leak will lead to the wastage of 640 liters of water in a month.

- Reuse the soapy water of washings form clothes for washing off the courtyards, driveways, etc.
- Water the plants in your kitchen-garden and the lawns in the evening when evaporation losses are minimum. Never water the plants in mid-day.
- Use drip irrigation and sprinkling irrigation to improve irrigation efficiency and reduce evaporation.
- Install a small system to capture rain water and collect normally wasted used water from sinks, cloth-washers, bath tubs, etc. which can be used for watering the plants.
- Build rain water harvesting system in your house. Even the President of India is doing this.

Mix Cycles

Unlike energy, the chemical elements tend to circulate in the biosphere in characteristic paths from environment to organisms and back to the environment.

Since organic substances are in limited supply in available form, their cyclic use and reuse constitutes an important part of the ecosystems. These more or less circular paths are known as biogeochemical cycles. Thus biogeochemical cycle may be defined as 'the more or less circular path which brings about the circulation of chemical elements, including all essential elements from the environment to the organisms and back to the environment.

Types of Biogeochemical Cycles: The biogeochemical cycles are of two types:

(1) The gaseous cycles in which the reservoir is in the atmosphere or hydrosphere and

(2) The mineral and sedimentary cycles, in which the reservoir is in the lithosphere, i.e. earth's crust.

Cycles of Gases

Oxygen Cycle: Atmospheric oxygen enters in the living world as a respiratory gas. The function of oxygen in respiration is to combine with hydrogen. Thes resulting water becomes part of the general water content of the living matter, as a source of element-hydrogen and oxygen. The structural oxygen remains in the organisms until decay and after death returns to the environment.

However, it usually returns not in the form of free atmospheric oxygen but as either water or carbon dioxide. The global oxygen cycle is interlinked closely with the global water and CO_2 cycles. A third possible fate of water within organisms (green plants) is its use as a raw material in photosynthesis, where it is split apart into hydrogen and oxygen. The hydrogen participates in food manufacture and oxygen as a by product comes out m the atmosphere.

Atmospheric oxygen is also the source of ozone (O_3) layer which envelops the earth at an altitude of some 10 miles. This layer prevents a great deal of the high-energy relation of the sun from reaching the earth's surface and so it affects plants metabolism indirectly by shedding potential lethal rays.

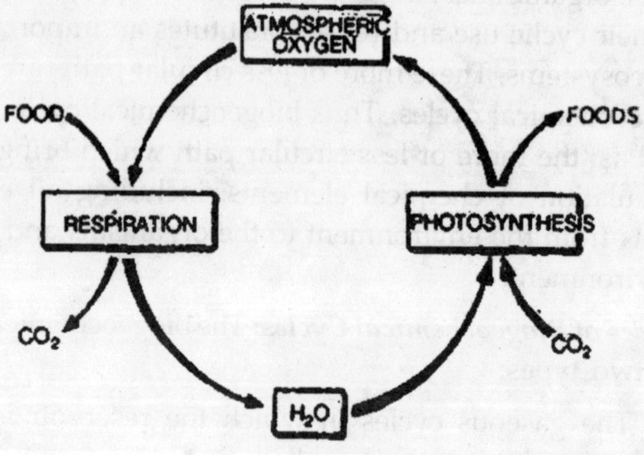

The Oxygen cycle.

Cycle of CO_2

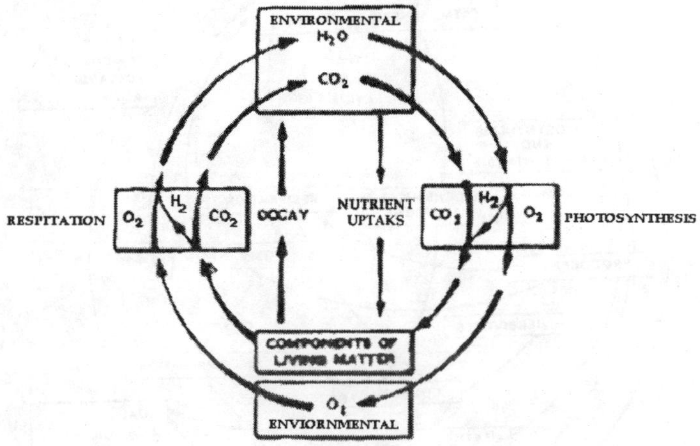

The inter-relation of the oxygen, water and carbon cycles.

Atmospheric carbon dioxide is virtually the exclusive carbon source and with water one of the two major oxygen sources for the construction of living matter.

The gas enters the living world through photosynthesis, in which it is a fundamental raw material. Photosynthesis incorporates CO_2 into organic substances and these are partly used in the construction of more living matter.

The CO_2 content of the atmosphere is replenished not only through biological combustion or respiration, but also through non-living combustion of real fires.

In the first case CO_2 is a by-product and is returned to the environment immediately; in the second it is a decay product returned after death.

The release of CO_2 into the air in forest fibres and in the burning of industrial fuels actually represents a long delayed completion of carbon cycle.

The combustible substances in wood, coal, oil and natural gas, all are organic compounds that are manufactured through photosynthesis.

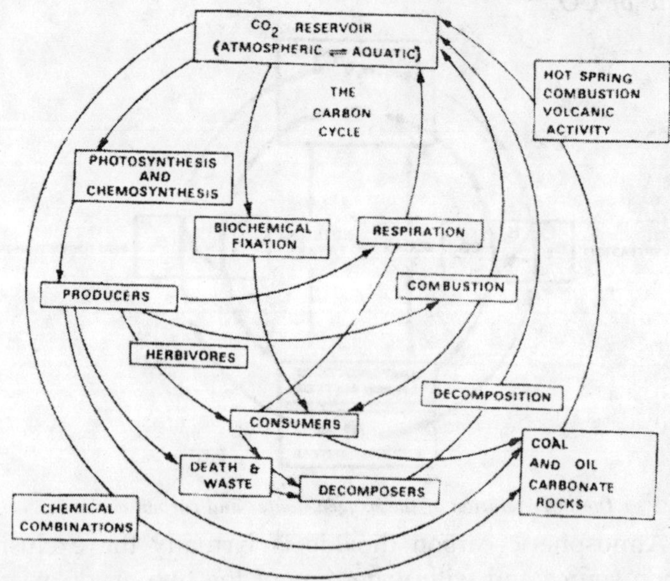

Carbon cycle.

Cycle of Nitrogen: Protein is the main part of protoplasm and nitrogen is an essential constituent of protein. Animals for their very existence and growth require large quantities of nitrogen. Though the atmosphere contains about 79% of nitrogen in the gaseous state, yet animals cannot utilise it unless it is converted into some usable form.

Animals take in nitrogen either in inorganic form as ammonia, nitrates and nitrites or in organic form as urea, proteins and nucleic acids. Thus atmospheric nitrogen can be used by animals only after it has been fixed as one of the inorganic substances. This fixation takes place both by biological as well as physio-chemical processes.

It was CAVENDISH who first noticed the increase in the quantities of nitrate in the soil after rain and thundershowers and explained, that during lightning union of nitrogen and oxygen occurs.

SCHONBEIN explained that small amounts of nitrogen

combine with hydrogen during evaporation to produce ammonium nitrate.

BERTHOLET demonstrated that small discharges of electricity cause the fixation of nitrogen with organic compounds. However, the fixation of nitrogen by these physio-chemical methods are very negligible and the only way of fixing of a major portion of nitrogen is through biological process.

The fixation of nitrogen by biological process is done by free living bacteria, symbiotic bacteria living in leguminous plants and by the blue green algae. Among the free living soil bacteria important one are the aerobic Azotobacter and the anaerobic Clostridium.

Azotobacter has a world-wide distribution and found in soil having a ph of 6 to 7. It has a high respiratory rate. The anaerobic form Clostridium is found in neutral soil. Both the genera produce ammonia as the first step.

Another group of bacteria break the large protein molecules into small nitrogen compounds that can be utilised by roots. It is these bacteria that help in the recovery of nitrogen compounds that are locked in the tissues of plants and animals in the normal process. This breakdown takes place in two parts:

(a) *Ammonification:* In this process ammonia is produced from the organic material of plants and animals. The bacteria that are involved in the process are Bacillus ramosus, B. mycoides and B. proteusvulgaris.

(b) *Nitrification:* Certain other bacteria transform ammonia into nitrates. This takes place in two steps. Nitrosomonas and Nitrococcus bacteria oxidise ammonia to nitrites. These nitrites are again oxidised by the bacteria Nitrobacter to nitrates. The action of these nitrifying bacteria are influenced by environmental and chemical nature of the soil. Both the types of bacteria work well in an alkaline medium.

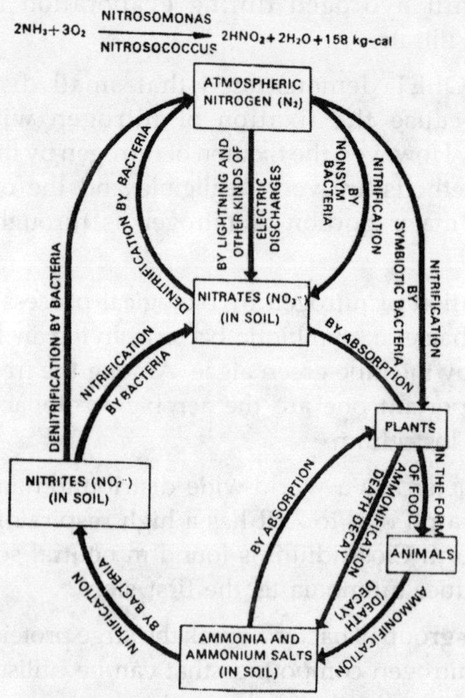

The Nitrogen cycle

The symbiotic bacteria Rhizobium radicicolum lives on the roots of the leguminous plants. The secretion of the plant roots help in multiply the bacteria. The bacterial excretions dissolve the cellulose of the root hairs, move in and bring about the rapid multiplication of cell. It is these cells that form the central mass of the root nodules.

The bacteria now change their shape from rod shape to bacteriod condition and start fixing nitrogen. The leguminous plant's thus fix nitrogen with the help of bacteria in agricultural land.

There are also non-leguminous nodule bearing plants which also do a similar job. They increase the nitrogen content of wide lands. Important non-leguminous plants which do nitrogen fixation are Alnus, Myrica, Ceanothurs and Shepherdia.

Blue green algae are also an important group of nitrogen fixers, Nostoc and Calothrix which are found in both aquatic and terrestrial habital contribute much in fixing nitrogen. These algae require minimum living conditions and are found to thrive in rice fields.

Cycling of Nitrogen: The main sources of nitrogen for plants and animals under natural conditions are that fixed by biological means in the form of nitrates from atmosphere, inorganic nitrogen brought by rain waters that are fixed by lightening ammonia absorbed by plants from soil and nitrogen released from dead organic material.

Plants utilise the nitrates and ammonia and covert them into amino acids. The consumers take these amino acids and convert them into different types of proteins. Animals also release some nitrogen by excretion as urea. Plants and animals after death and decay are broken down by bacteria and fungi and ammonia is released.

Ammonia can be utilised by plants to form nitrates and nitrites or can find their way to the atmosphere. Nitrates and nitrites can be carried away by rivers to lakes and also to the sea. There are certain bacteria called the denitrifying bacteria. Important among them is Bacterium denitrificans. They use nitrates as nutrients and ultimately convert them to molecular nitrogen. As this gas escapes into the atmosphere global cycle becomes complete.The nitrogen cycle is also affected by human intrusion. Grassland cultivation and the removal of harvested crops reduce the nitrogen content. Felling of trees for timber brings out a heavy outflow of nitrogen from the forest ecosystem.

An addition of commercial fertilizers increases the quantum of nitrogen in the soil. Cultivation in general increases the nitrogen content of the soil. If leguminous plants are cultivated then the increase is many times. On the whole the nitrogen cycle has to be carefully studied and any break in the ecosystem must be avoided.

Cycles of Sediments: In the sedimentary type of cycle, the major reservoir is the lithosphere from which the elements are released by weathering. The sedimentary types are best exemplified by phosphorous and sulphur. Both sulphur and phosphorous are very important factors, limiting or controlling the abundance of organisms. Both these elements have a tendency to stagnate as portion of the supply is lost as in deep sea sediments, and thereby become inaccessible to organisms and to continual cycling.

Cycle of Phosphorous: Phosphorous is an important and necessary constituent, of protoplasm as it is an essential constituent of nucleic acids, phospholipids and numerous phosphorylated compounds. The phosphorous tends to have been formed in past geological ages. These are gradually eroding, releasing phosphates to the ecosystems. Plants absorb these inorganic phosphates dissolved in water. These phosphates are transferred to animal consumers and decomposers as organic phosphates and subsequently is made available for recycling via mineralizing decomposition. Much phosphate is leached into the sea, where part of it is deposited in the shallow sediments and part of it is lost to the deep sediments.

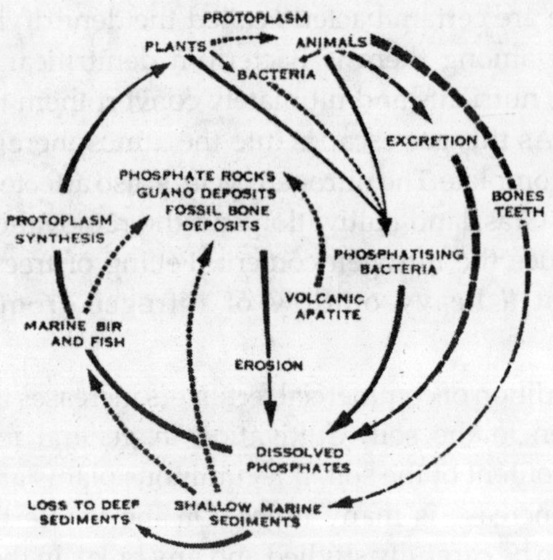

The phosphorous cycle.

Cycle of Sulphur: Sulphur is an essential constituent of certain amino acids. Not as much sulphur is required by the organisms as nitrogen and phosphorus, nor it is an often limiting to the growth of plants and animals.

Only a few organisms obtain their sulphur requirements in such organic forms as amino acids and cystine. The inorganic sulphate is the major source of biologically significant sulphur. Most of the biologically incorporated sulphur is mineralised by bacteria and fungi. Under anaerobic condition, however, hydrogen sulphide and some others are reduced direct to sulphides by bacteria as Escherichia and Proteus. Some organic sulphur enters the atmosphere as SO_2 by incomplete combustion of fossil fuels. Inorganic sulphur precipitates out as sulphate. Since sulphate is readily soluble in water, it serves as a source of elemental sulphur in many ecosystems. Sulphate is also reduced under anaerobic conditions to elemental sulphur or to sulphides by heterotrophic bacteria such as Desulfavibrio. The sulphate reducing anaerobic bacteria are heterotrophic, using the sulphate as a hydrogen acceptor in metabolic oxidation in a manner comparable to the use of nitrite and nitrate by denitrifying bacteria.

Colourless sulphur bacteria Beggiaota oxidise hydrogen sulphide to elemental sulphur and Thiobacillus oxidise it to sulphide. Certain species of Thiobacillus oxidise sulphide to sulphur and still others oxidise sulphur to sulphate. In some species oxidation process takes place in the presence of oxygen while in others presence or absence of oxygen hardly matters. These latter bacteria include the green and purple photosynthetic bacteria that use the hydrogen of H_2S as the oxygen acceptor in reducing CO_2; the green bacteria are able to oxidise sulphide to elemental sulphur, whereas the purple bacteria carry the oxidation to the sulphate stage.

The sedimentary aspect of the cycle involves the precipitation of sulphur in the presence of iron under anaerobic conditions. Ferrous sulphide is insoluble in neutral or alkaline water, and consequently sulphur has the potential for being bound up under these conditions to the limits of the iron present. Because of the thermodynamics of the ferrous sulphide system, other nutrients important to biological system can also get trapped for varying periods of time-copper, cadmium, zinc and cobalt are the examples of such nutrients. On the contrary, the very binding of these iron compounds allows the conversion of phosphorous from insoluble to soluble and thus makes it accessible. The sulphur cycle provides an excellent example of the interaction and regulation that exists between different mineral cycles.

Functions of Atmosphere

Atmosphere is also subjected to physical cycles by the sun and the spin of the earth. Warm equatorial air rises, cool polar air sinks and the axial rotation of earth shifts air mass laterally. Like the water circulation, global air circulation also influences climatic conditions substantially. Equally significant to animals are the chemical cycles of the atmosphere. Air consists mainly of O_2 (20%), CO_2 (0.3%), and N_2 (79%). Water and minute traces of inert gases (He, Ne, Kr and Ar) are also present in the atmosphere. Except inert gases all the components of air serve as raw materials and each plays a conspicuous role in the global cycle.

Functions of Hydrosphere

Water represents the most abundant mineral of the plant. It covers some 70% of earth's surface entirely, and it is a major constituent of the lithosphere and the atmosphere. Water is also the most abundant component of living matter.

The basic cycle of water in the environment is quite familiar. Solar energy evaporates water from the hydrosphere into the atmosphere. Later, cooling and condensation of the vapours at higher altitudes produce clouds-precipitation as rain or snow. then returns the water to the hydrosphere. This is the most massive process of any kind on earth consuming more energy and moving more material than any other process.

Aquatic animals obtain water from their environment and return a part of it throughout excretion during their life time. After death, the remaining water is returned through decay. Terrestrial organisms, i.e. plants and animals are interposed more extensively in the global cycle of water.

These organisms draw water from the reservoirs present in the soil such as rivers, streams, lakes, ponds, etc. Plants and animals move this water through their bodies where some of it is retained. The rest is excreted, partly as liquid water back into the hydrosphere but more particularly as water vapour into the atmosphere. Actually terrestrial organisms accelerate the global water cycle. After the death of terrestrial organisms, any liquid water present in their body returns to the hydrosphere through decay.

Though a prime nutrient, water also influences animals through its effects on climate and weather, both on sea and on land. In the ocean, water warmed in the tropics becomes light and rises to the surface, whereas cold polar water sinks. These up and down displacements of water along with the driving force of the winds result in ocean currents. Such currents influence climatic conditions not only within the sea but also in the air and on land. As a consequence of its thermal properties, water has another climatic effect. Of all liquids, water is one of the slowest to heat or cool and stores a very large amount of thermal energy.

Thus oceans have become huge reservoirs of solar heat. The result is the sea-air chilled at night becomes less cold because of heat radiation from water warmed by day. Conversely, sea-air warmed during the day, becomes less hot because of heat absorption by water cooled at night. Warm and cool inshore winds then produce seasonal summer and winter patterns.

Long-range global climates are influenced by the relative amount of water locked into polar ice. Temperature variations averaging only a few degrees produce over the years by geophysical changes is suffice for the advance or retreat of polar ice. At present the earth is slowly emerging from the last ice age, and as polar ice is now melting, water levels are rising and coast lines are gradually being submerged.

All these changes in the hydrosphere have profound impact on animals. By influencing temperature, humidity, amount of precipitation, winds, waves and indeed the very presence or absence of water in given localities, they play a major role m determining what kinds and amounts of animals can live in such localities.

Functions of Lithosphere

Lithosphere not only supplies most mineral nutrients to all animals but it also forms the sand base of soil, specifically, required by terrestrial plants and by numerous subterranean animals.

Like the world's water, the rocky substance of the earth surface moves in a gigantic cycle. But here the rate of circulation is measured in thousands and millions of years. One segment of this global mineral cycle is diastrophism, the vertical uplifting of large tracts of earth's crust through a variety of geological forces. Major parts of continents or indeed whole continents may undergo such slow diastrophic movements.

The most striking example is mountain building. Presently the youngest and highest mountains are the Himalayas, the Andes and the Alps. All of them were thrown up during the laramid revolution, some 70 million years ago. The earth surface in this region is not completely settled even now.

The appearance of a high mountain barrier is likely to interfere drastically with continental air circulations. The moisture laden oceanwinds may no longer be able to pass across the barrier. Rain will continue to fall in the region between ocean and mountain, which will become lush and fertile. But the region beyond the mountain will be arid and desert conditions are likely to develop. Second segment of the global lithospheric cycle involves gradation, the lowering of high land and the levelling of mountains. These changes result in part from erosion and dissolution of rocks and in part from actual geologic sinking of land by the hydrosphere and the atmosphere.

The physical forces of gradations are chemical forces, which are particularly important to animals. The chemical processes of decay are major erosive factors. They contribute to the breaking of large stones into smaller ones and then into tiny sand grains and microscopic rock fragments. Chemical erosion thus play a major role in the formation of soil particles.

Second, whenever water comes in contact with a rock it dissolves small quantities of it and so acquires any mineral content. Dissolved soil minerals eventually drain into rivers and rivers drain into the ocean. Hence as the lithosphere is slowly being denuded of mineral compounds, the hydrosphere fills with them. It was partly because of it that the early seas on earth acquired their original saltness. Plants in the sea freely use the minerals as metabolites.

The death of marine organisms helps to complete the global mineral cycle. Many protists and animals use mineral nutrients in the construction of protective shells and supporting bones. After these organisms die, their mineral components rain down to the sea floor where, as pointed out earlier, the accumulating layers of ooze may ultimately compress into rock.

The global lithospheric cycle becomes complete when such a section of sea bottom or low lying land is subjected to new diastrophic forces. High ground or mountains are thereby regenerated and parts that were seafloor originally may be thrust up as new land in the process.

The lithospheric cycle supplies many more types of minerals than normally plants require. The minerals required and used by the plants are nitrates, phosphates, chlorides, carbonates, sulphates, and sodium, potassium, calcium, manganese, magnesium, copper and iron. Despite the dense cover of plant life carpeting the earth, most of the minerals are still well in excess of currently required amounts. However, because of the uninterrupted global growth of plants for million of years, a few key minerals now tend to be in relatively short supply. In particular phosphates and nitrates have become significant limiting factors. Thus agriculture soils are artificially enriched with fertilizers, largely phosphates and nitrated.

The Biosphere

The part of the earth containing living organisms is known as 'biosphere'. It can also be defined as 'the part of the earth in which the ecosystem exists'. The physical portion of the biosphere consists of following three subdivisions:

(1) Hydrosphere, which includes all liquid components namely that water in oceans, lakes, rivers and on land;

(2) Lithosphere, which comprises the solid components, the rocky substance of the continents and

(3) Atmosphere, the gaseous mantle around the hydrosphere and lithosphere.

Plants require inorganic metabolites from each of the subdivisions of the biosphere. The hydrosphere supplies liquid water; the lithosphere supplies all other minerals; and the atmosphere supplies oxygen, intergang and carbon dioxide. Together, these inorganic materials provide all the chemical elements needed in the construction and maintenance of living matter.

3

Geographical Factors

The living organisms in a given region are collectively termed as biota of that region. The plants of a given region are termed the flora and the animals of a particular region are termed as fauna.

The study of the distribution of biota is called Biogeography; those of plants phytogeography and of animals zoogeography.

Status of Climate

Australian Region is partly tropical and partly temperate. New Guinea is tropical and is mostly covered with rain forest; the Northern part of Australia is tropical but most of the interior is arid. Tasmania is cool and temperate.

Fauna: The vertebrate fauna of Australian Region is very thin. It lacks both the variety as well as the number of families, but it has several unique genera. The most peculiar feature is the absence of higher placental mammals and it contains very many primitive forms. Marsupials and monotremes are found only in this region. The vertebrate fauna of Australian Region is very thin. It lacks both the variety as well as the number of families, but it has several unique genera. The most peculiar feature is the absence of higher placental mammals and it contains very many primitive forms. Marsupials and Monotremes are found only in this region.

Fishes: Australian freshwater fishes are Osteoglossids and Neoceratodus but both have restricted distribution.

Amphibians: Amphibians are few. The common toads and tailed amphibians are absent. The Australian representatives are Frogs (Hyla and Rana).

Reptiles: Australian reptiles are moderately varied and only two families are exclusive. Crocodiles, Turtles, Geckos, Skinks, Varanus, Typhlops, Python, some Colubrids occur in New Guinea and part of Australia. Python and biting tiger snakes are abundant.

Birds: Bird fauna is abundant having 58 families. About 44 families are more or less widely distributed over the world. Two families are divided between Australia and Oriental regions, but 12 are exclusive forms.

Trogon and Kingfishes, Hawks, Cuckoos, Pigeons and Parrots and very numerous. The Pigeons and Parrots have reached their maximum diversity.

There are three exclusive families of Parrots, Cuckoos, Loris and Pigmy parrots. Frogmouths, Wood-swallows, Flower peckers and Megapodes are shared with Oriental Region, but Pheasant, Finches, Barbets and Woodpeckers are absent. Total 10, families of birds are exclusive. These are:

(i) Cassowaries
(ii) Emus
(iii) Honey suckers
(iv) Lyre birds
(v) Bower birds
(vi) Legendary birds of paradise
(vii) Megapodes

(viii) Owlet frogmouths
(ix) Flower peckers
(x) Bell Magpies and
(xi) Scrub birds

Mammals: The mammalian fauna of Australian Region is characterised by the complete absence of higher eutherian mammals. There are a few monotremes but several families of marsupials. Echidna and Omithorhynchus are monotremes found in Australian region only.

Dasyurus, Perameles, Opossum, Bandicoots, Wombats and Phascolomidae are the six marsupial families. There are 6 families of insectivorous Bats and few fruit Bats. Mice, Australian dog and European rabbit and hare have been introduced into this region. In addition, Phalanger family is represented by Opossums, Squirrels, Flying Phalangers.

Thus the Australian fauna is remarkable for its poverty of freshwater fish, amphibians and reptiles and for the uniqueness of its mammals. Australian Region has little in common with Ethiopian Region but shows close affinity with Oriental Region.

Vegetation being Key Functioning

***Forest Vegetation*:** CHAMPION and SETH (1968) have recognised the following sixteen types of forests in India:

Significance of Moist Tropical Forests

1. *Tropical Wet Evergreen Forests:* These forests are found in Western Ghats, Assam, Cachar, parts of Bengal, Mysore and in Andamans. These are climatic climax forests with very dense growth of tall trees (more than 45 metres high). The shrubs, lianas, climbers and epiphytes are abundant. However, grasses and herbs are rare on the forest floor. The rainfall exceeds 250 cm. in these regions. The characteristic flora of these forests includes Dipterocarpus indica, Hopea, Artocarpus, Mangifera, Emblica, Michelia, Ervatamia, Ixora, etc.

2. *Tropical Moist Semi-evergreen Forests:* These forests are found along the Western Ghats, in part of Upper Assam and Orissa. The rainfall is usually more than 200 cm. per year. These forests consist of dense and tall (25 to 35 metres) trees along with shrubs. In these forests, deciduous species like Terminalia, Tetrameles and Shorea grow intermixed with the evergreen species like Artocarpus, Michelia and Eugenia.
3. *Tropical Moist Deciduous Forests:* These forests are distributed in a narrow belt along the foot of the Himalayas, and also on the eastern side of Western Ghats, Chhota Nagpur and Khasi Hills. These have average rainfall of 150-200 cm. The trees are deciduous and remain leafless for one or two months. The common species are teak (Tectona grandis) in the south and sal (Shorea robusta) in the north. Other common species are Dalbergia, Cedrela, Salmalia, Albizzia, Terminalia, Cordia, Bombusa, Melia, Dillenia, Eugenia, Dendrocalamus, etc.
4. *Littoral and Swamp Forests:* These comprise mostly evergreen species and occur along the sea coast in wet and marshy areas and also in deltas of larger rivers on eastern coast. In saline swamps, mangroves are chiefly composed of Rhizophora, Bruguiera, Ceriops, Nipa, etc. In less saline areas Phoenix, Ipomoea, Phragmites, Casuarina, etc. are commonly found.

Significance of Dry Tropical Forests

5. *Tropical Dry Deciduous Forests:* These are distributed almost throughout the country except Kashmir, beyond Bengal, Rajasthan and Western Ghats. These occur in areas with annual rainfall averaging from 75-125 cm. and with a dry season of about 6 months. The trees are small. Shrubs are abundant. Shorea robusta (in north), Tectonagrandis (in south), Anogeissus, Tenninalia, Semecarpus, Carissa, Emblica, Acacia, Ziryphus, Dalbergia, etc. are found growing in these forests.

6. *Tropical Dry Deciduous Forests of South-East Deccan Region:* In some parts of Tamil Nadu and the coast of Karnataka ever green forests of short (10-15 metres) but dense trees are found. The anrual rainfall is relatively small. There are no bamboos but grasses are abundant. The common flora of this region is Memecylon, Maba, Pavetta, Feronia, Ixora, Sterculia, etc.
7. *Tropical Thorn Forests:* These are distributed in Western Rajasthan, parts of Punjab, Bundelkhand, Gujarat, Maharashtra, Madhya Pradesh and Tamil Nadu. The trees are short (8-10 meters in height), sparsely distributed, mostly thorny. Plants remain leafless throughout the year. During rains, grasses and herbs remain leafless throughout the year. During rains, grasses and herbs become abundant. The common plants of these forests are Acacia, Prosopis, Capparis, Albizzia, Ziryphus, Euphorbia, Calotropis, Madhuca, Tephrosia, Crotalaria, etc.

Significance of Subtropical Forests

8. *Subtropical Broad-leaved Hill Forests:* These forests have predominantly evergreen species, and occur in relatively moist areas as lower slopes of eastern Himalayas, Bengal, Assam and hill ranges of Khasi, Nilgiri, Mahabaleshwar, etc. In the north occur Quercus, Castanopsis, Schima and some temperate species. In eastern Himalayas, due to higher humidity, bamboos, several epiphytes including orchids and ferns become abundant. The common trees of south are Eugenia, Randia, Ficus, Populus, Canthium, Mangifera and the climbers are Piper, Gnetum, Smilax, etc.
9. *Subtropical Pine Forests:* In Western Himalayas and Khasi Jayantia hills of Assam, open forests are composed mainly of pine species like Pinus taxburghi and P. khasya.

10. *Subtropical Dry Evergreen Forests:* Lower elevations of Himalayas have forests with thorny and small-leaved evergreen species. These areas are characterised by low rainfall and low temperature. The common species are Acacia modesta, Dodonea viscosa, Olea cuspidata, etc.

Significance of Temperate Forests

These forests are found at a height of 1800 to 3800 metres in the Himalayas.

11. *Montane Wet Temperate Forests:* These forests are found in eastern Himalayas with high rainfall and also in Nilgiris in the south. The forests in the south are evergreen and are known as Sholas. The trees are 15 to 20 metres high with a dense growth. Epiphytes are abundant. The common plant species are Rhododendron nilagircum, Hopea, Artocarpus hirsuta, Salmalia, etc.

    In the north, the west temperate forests extend from eastern Nepal to Assam at a height of 1800 to 2900 metres. These forests are evergreen, dense and high (upto 25 metres). The common plant species include Quercus, Acer, Prunus, Ulmus, Begonia, Loranthus, etc.

    In the western Himalayas, Quercus species dominate all over. Conifers like Abies, Cedrus, Picea, etc., and are also widely distributed.

12. *Himalayan Moist Temperate Forests:* Central and Western Himalayas have conifers and oaks at an altitude of 1500 to 3000 metres. The trees are tall (upto 45 metres), but undergrowth is thin and deciduous.

    These forests have plants like Cotoneaster, Berberis, Spiraea, Quercus, Cedrus, Pinus, Picea, Abies, Tsuga, etc.

13. *Himalayan Dry Temperate Forests:* These forests occur in a narrow belt in the western Himalayas extending from

parts of U.P. through Himachal Pradesh and Punjab to Kashmir. They are evergreen forests with scrub undergrowth. Oaks and conifers are dominant. In western drier areas Pinus gerardiana and Quercus ilex are common, while in the eastern moist zone Abies, Picea, Larix, Juniperus are found.

Significance of Alpine Forests

14. *Sub-alpine Forests:* These evergreen forests are open and occur throughout the Himalayas at high altitudes (above 3000 metres upto tree line). Conifers like Abies spectabilis, broad-leaved species like Butea and Rhododendron, and shrubs like Cotoneaster, Rose, Lonicera, Smilax, etc. are found in abundance in these forests.
15. *Moist-alpine Scrub:* Above the timber line upto the height of 5500 metres in the Himalayas, are found dwarf, evergreen, shrubby growth of conifers (Junipers) and broad-leaved species like Rhododendron.
16. *Dry-Alpine Scrub:* In areas with annual rainfall below 35 cm. are found open xerophytec scrubs like Juniperus, Caragana, Eurota, Salix, etc. are found upto the height of 5500 metres.

Significance of Grasslands

Following three types of grass lands occur in India:

(i) Xerophilous, found in dry regions of north-west parts of the country in semi-desert conditions.

(ii) Mesophilous, are extensive grass fields or savannas found in moist deciduous forests of U.P.

(iii) Hygrophilous, also called wet savannas.

Significance of Floristic Areas

The Indian subcontinent is characterised by a variety of climates and flora.

It has been divided into the following nine floristic regions for the study of flora:

1. *Western Himalayas:* Extends from Kumaon to Kashmir. The annual rainfall is about 200 cm. It has been divided into three zones of vegetation corresponding to three climatic zones:
   (i) *Submontane Zone:* It extends upto 1500 metres in height and comprises of Shiwalik ranges. The forests are tropical and subtropical having trees Shorea robusta, Dalbergia sisoo, Cedrelatoona, Ficus glomerata, Eugeria jambolano, Acacia catechu, Ziryphus, Butea monosperma and thorny euphorbias.
   (ii) *Temperate Zone:* It extends from 1500-3500 metres in altitude. The dominant plant species are Quercus, Acer, Ulmus, Rhododendron, Betula, Salix, Prunus, Pinus, Cedrus and Taxus.
   (iii) *Alpine Zone:* This zone extends from 3500-4500 metres. It is characterised by alpine forest vegetation with scrub and meadows. Abies, Betula, Juniperus and Rhododendron are the common species of trees found in this region. Herbs like Primula, Potentilla, Polygonum, Geranium, Sexifraga, etc. occur near the snow line.
2. *Eastern Himalayas:* It comprises Sikkim and NEFA regions and is characterised by more rainfall, less snow and modest to high temperature. This region has been divided into three subregions:
   (i) *Tropical Zone:* It extends upto 1800 metres in height and has tropical and semi-evergreen or moist deciduous forests. Plants like Acacia catechu, Delbergia sisoo, Shorea robusta, Cedrela, Dendrocalamus (bamboo) and Albizzia are commonly found in this zone.
   (ii) *Temperate Zone:* This zone extends between

1800 metres and 3800 metres. It has typical montane temperate forests which are dominated by oaks like Michelia, Quercus, Pyrus, Eugeria at lower levels. Conifers like Juniperus, Cryptomeria, Abies, Pinus, Larix, Rhododendron, bamboo, etc. are found in higher and cooler regions.

(iii) *Alpine Zone:* Beyond the temperate zone, extends alpine zone upto the height of 5000 metres. It has alpine vegetation but trees like Juniperus and Rhododendron are also found in this region.

3. *Indus Plains:* This zone includes arid and semi-arid regions of Punjab, Rajasthan, Kutch, parts of Gujarat and Haryana. The average rainfall is less than 70 cm. Tropical thorn forests are found in semi-arid regions but the arid zone is typical desert.

   The plants growing in this region are xerophytes like Acacia nelotica, Prosopissp, Salvadora, Tecomella, Capparis, Tamarix, Zizyphus, Calotropis, Saccharum, Conchrus and Euphorbia.

4. *Gangetic Plains:* This region comprises of U.P., Bihar, Bengal and part of Orissa. This region is characterised by the most fertile land in the world. Vegetation in mainly of tropical moist and deciduous forest type.

   The common plants growing in the region are Dalbergia sisoo, Acacia nelotica, Saccharum munja, Butea monosperma, Madhuca indica (mahua), Buchanania lanzan (chiraunji), Diospyros melanoxylon (tendu) Cordi myxa (lisora), Acacia catechu (khair), Azadirachta indica (neem), Ficus bengalensis (bergad), Ficus refgiosa (pipal), Mangifera indica (mango) and weeds and grasses like Xanthium, Cassia, Argemone, Amaranthus, etc. In Gangetic delta of sunderban (South Bengal) mangrove vegetation is common.

5. *Central India:* It comprises of M.P., parts of Orissa and Gujarat. The average rainfall is 150-200 cm. Vegetation

is primarily thorny, mixed deciduous and teak. The dominant plants of this region are Tectonagrandis, Madhuca, Diospyros, Butea, Dalbergia, Terminalia, Carissa, Ziryphus, Acacia, Mangifera, etc.

6. *Malabar (West Coast):* This region includes western coast of India from Gujarat to Cape Comorin. This zone has moderate to high rainfall. The forests are tropical evergreen in extreme west, semievergreen towards interior regions, subtropical temperate evergreen forests in Nilgiris and mangroves near the coasts of Bombay and Kerala.

7. *Deccan Plateau:* This region extends all over peninsular India (i.e., Andhra Pradesh, Tamil Nadu and Karnataka) the average rainfall is 100 cm. The central hilly plateau has tropical dry deciduous forests of Boswellia serrata, Tectona grandis and Hardwickia pinnata, and the low eastern dry. Coromandal Coast has tropical dry evergreen forests of Santalum album (chandan), Cedrela toona.

   The plants like Acacia, Prosopis, Euphorbia, Capparis, Phyllanthus, are also found growing here.

8. *Assam:* This region is characterised by heavy rainfall. The vegetation is either dense evergreen or sub-tropical. In evergreen forests are found trees like Dipterocarpus macrocarpus, Mesua ferrea, Shorea robusta, Ficus elastica, etc., bamboos as Bambusa padida, Dendrocalomus hamiltonii, etc., grasses like Imperata cylindrica, Saccharum, sp., Themeda sp., insectivorous plants like Nepenthes and epiphytes (ferns and orchids) are found growing in this region.

9. *Andmans:* This region has varied type of vegetation. It is characterised by mangroves and beech forest at its coasts and evergreen forests of tall trees in the interior, Important plant species of this island are Rhizophora, Mimusops, Calophyllum, Lagerstroemia, etc.

Zoogeography of Indian Subregion

Geographic Limits: BLANFORD in 1901 discussed the geographical distribution of animals in British India, which included not only the territory of present India, but also Pakistan, Kashmir, Gilgit, Laddakh, Nepal, Sikkim, Bhutan and other Himalayan states, Garo, Khasi, Naga Hills and Manipur; Burma and also Andaman and Nicobar Islands. At present Laccadive and Maldive islands have also been added to this.

SMALLER AREAS IN FLORA

Owing to the varied climatic and geographic conditions fauna of Indian Subregion is also very heterogenous and further divisions of Indian Subregion is not an easy task. The fauna of different divisions exhibits affinities. Its division has been attempted by a number of scientists.

JORDAN (1862) divided it on the basis of distribution of birds; GUNTHER (1864) on reptiles; BLANFORD (1876) on molluscs and WALLACE (1876) on the basis of distribution of all animal groups. B. PRASAD (1921) subdivided Indian Subregion into five subdivisions, but recently MAHENDRA (1942) has divided it into ten subdivisions:

1. The arid and semi-arid province of North India
2. The Western Himalayas
3. Southern Burmese province
4. Transgangetic province
5. Gangetic plain and adjacent part as far south as 20° lat.
6. South India below 20° lat. excluding Travancore
7. Travancore province
8. Ceylon
9. Andaman Islands
10. Nicobar Islands.

1. *Arid or Semi-arid Province of North India:* It embraces

West Pakistan, Punjab, Western Rajasthan and Cutch. Fauna is not very remarkable with no endemic forms. Amphibians are few represented with cosmopolitan genera Rana and Bufo only.

2. *Western Himalayas:* It comprises of Kashmir, Shimla, Kumaon and Gharwal districts, adjacent mountainous regions of Western Tibet and Punjab. The zone is poor in amphibians and reptiles. Fauna is markedly different and includes a number of endemic species, *e.g.* Gymnodactylus fasciolatus, G. lawderanus, Japalura major and J. Kumaonensis, Phrynocephalus theobaldi and Leiolopisma indacenes.
3. *Transgangetic Province:* It includes northern points of Bihar, Assam and Burma north of 20° lat. and Nepal, Himalayas east of Nepal, Sikkim and Bhutan. This Subdivision has a characteristic reptilian fauna with a number of endemic genera and species.
4. *Southern Burmese Province:* It includes southern part of Burma. It includes certain characteristic snake species and some endemic amphibians. Gymnodactylus oldhamin, Riopa anguina, Bungarus flaviceps, Doliophis intestinalis are characteristic snake subspecies of this region.
5. *Gangetic Plain and Adjacent Region as far South 20° lat:* The boundaries of this subdivision are not very precise. The subdivision is characterised by the extreme poverty of fauna and has no endemic species. There is complete absence of House gecko, Gavialis and almost all freshwater Turtles.
6. *South India Below 20° lat. Excluding Travancore Cochin:* The fauna of this subdivision is very characteristic including Riopa lineala, Barkudia insularis, Gymnodactylus dekkanensis and Hemidactylus subtriedrus.
7. *Travancore:* It consists of hilly region of the country south of lat. 12 or 13 on the west and south of the

Coleroon river in the east. Its fauna is rich in endemic species.

8. *Ceylon:* It forms a distinct subdivision of Oriental Geographical realm and had clear affinities to Travancore subdivision.
9. *Andaman Islands:* These are situated in the Bay of Bengal and have very characteristic fauna with clear affinities to Burmese subdivision on one hand and the Nicobar Islands on the other hand. The endemic species are Gymnodactylus rubidus, Phelsuma andamanesis, Calotes anda, manesis, Typhlop andamanesis and Bioga andamanesis.
10. *Nicobar Islands:* Its fauna has affinities with Andaman Islands, and contains forms like Calotes mystaceus, Mabuya andamanesis, Goniocephalus suberistatus. The endemic species are few in number and the fauna of Nicobar appears to be derived recently from Andaman Islands.

Functions of Neotropical Region

Geographic Limits: It embraces South America, Central America, Tropical low land of South Mexico and West Indies.

Climate: Neotropical region presents tropical conditions except Southern part of South America which constitutes South Temperate zone. It is noted for luxuriant forests and complete absence of desert.

Extensive rain forests of evergreen forests are found in Amazon Valley, tracts of dry forests and grassy plains extend in Savanna and Argentina, and subdesert areas are present in western South America. In west is seen long range of Andes which has high mountain forests, plateau land and gentle slopes. This suggests that vegetation is rich and complexly distributed.

Fauna: The fauna of Neotropical Region is both distinctive and varied. It is rich in endemic families of all the classes. In all it has 155 families of terrestrial vertebrates, of which nearly 39 are endemic. The rest are shared with Nearctic and other typical regions.

Fishes: Freshwater fishes are very numerous. But the notable feature is the complete absence of carp family (Cyprinidae). It is dominated by Characian fish, Cat-fish, Gymnotids (electric eels) and Trygonids. Species of electric eels (Gymnotids) are found in Amazon. One genera of lung fishes (Lepidosiren) and Gar-pike (Lepidosteus) are also found. The fishes are numerous and endemic but with general African relationship.

Amphibians: Amphibians are represented by fourteen families of Anurans of which the genera of the Pipa, Hyla, Bufo and Rana are the common ones. The four families are peculiar namely Hylaplesidae, Plectomantidae and Pipidae. Caecilians are represented by Siphonopsis and Rhinotrema. The urodeles are just few and the peculiar genera is Spelerpes.

Reptiles: The reptilian fauna is shared with Nearctic, Ethiopian and Oriental Regions.

(i) *Crocodiles and Turtles:* Crocodiles and its few species, Alligators, Turtles and tortoises are very common. The tortoises are represented by Dermatemys, Stoutotypus, Peltocephalus, Padocnemis, Hydromedusa and Chelys.

(ii) *Lizards:* This region is inhabited by fifteen families of lizards of which five are peculiar, namely Helodermidae, Anadiadae, Chirocolidae, Cercosauridae and Iphisiadae. Geckos and Iguanas are represented but families Varanidae, Lacertidae, Agamidae are entirely absent.

(iii) *Snakes:* Snakes are represented by Typhlops, Leptotyphlops, many Colubrids, Coral snakes, Boas, and Pit-vipers. The characteristic genera are Dromicus, Epicrates, Ungalia and Elops, etc.

Birds: The avian fauna is so striking and diverse that South America has been called the "Bird Continent". About 67 families are represented of which abut 23 families are restricted and two others are totally confined to this region. Rhea (American Ostrich) and Tinamus are endemic genera of flightless birds. Other notable features are complete absence of song birds from the continent and the quails are the only representatives of Pheasant family.

The common birds which have wide distribution are Herons, Storks, Ducks, Plovers, Pigeons, Parrots, Goatsuckers, Swifts, Woodpeckers, Swallows, Thrushes, Kingfishers, lbeses and Barets. The endemic families can be summarised as:

Pipridae, Todidae (todies), Momotidae (momot), Carcidae (curassows), Tinamidae (Tinamus), Aramidae, Eurypygidae (sun bitterns), Cariamidae, Bucconidae (puff birds), Steatornithidae (oil birds), Psophildae (trumpeters), Falamedidae (horned screamers), Contingidae (chaterers), Dendrocolaptidae (tree creepers), Rhamphastidae (toucans), Formicariidae (sand thrushes), Caerebidae (sugar birds), Phytotomidae (plant cutters), Galbulidae (jacanas), Opisthocomidae (hoazin), Rheldae *(Rhea),* Conophagidae and Thinocoridae.

Mammals: Mammalian fauna is also distinctive and varied. There are 32 families of mammals excluding bats, of which seven are widely distributed and ten or more endemics. Hedgehogs, Moles, Beavers, Hyenas, Bovids and native Horse are absent. But it contains Oppossums, Coenolestid marsupials, Anteaters, Sloths, Armadillos, Tapirs, New World Monkeys (Squirrels, Monkeys and Spider monkeys), Rabbits, Deers, Llamas, (camel-like forms), Squirrels, New World Porcupines, eleven families of rodents and five indigenous families of bats. The endemic families are enumerated as under:

1. Cebidae ⎤ New World Monkeys
2. Hapalidae ⎦
3. Chinchillidae–Chinchillas ⎤
4. Caviidae–Cavies ⎥ Rodents
5. Octodontidae ⎦

6. Echimyidae
7. Bradypodidae—Sloths
8. Dasypodidae—Armadillos
9. Myrmecophagidae—Anteaters
10. Didelphidae-Didelphis-Marsupial
11. Coenolestidae-Marsupial

Key Role of Subregions

The Neotropical Region has been divided into four Subregions:

1. Chillian Subregion
2. Brazilian Subregion
3. Mexican Subregion
4. Antilean Subregion

1. *Chillian Subregion:* It includes western coast of South America and embraces the summits of Andes of Peru and Bolivia. The region does not contain any peculiar family and the characteristic forms are Chinchillas, Llamas, Oil birds and Rheas.
2. *Brazilian Subregion:* It includes tropical forest region of South America terminating northwards at the Isthumus of Panama. It has evergreen jungles, open plains and pasture land. This subregion is characterised by varied and rich fauna. All the arboreal animals of Neotropical Region are confined to this subregion. American monkeys, Vampire bats, American porcupines, Sloths, Armadillos and Oppossums are found here. In addition to these some other forms like Tapirs, Cavies and Spiny mice, etc. are also found here.
3. *Mexican Subregion:* This part of Neotropical. Region north to Isthmus of Panama is represented by Mexican Subregion. It is in the form of an irregular neck of land which connects North America with South America. It

comprises of mostly the rocky mountains. The mud terrapins, Tapirs, Anguidae and Plethodontidae are its characteristic fauna.

4. *Antilean or West Indies subregion:* It comprises of Islands of West Indies except Tobago and Trinidad. This includes Cuba, Haiti, Jamaica, Porto Rico and chain of smaller islands-St. Croxis, Angulla, Barbuda, Antigua, Barbads and Grenada. Most of the islands are mountainous and rocky, covered by forests. Since this Subregion is wholly made up of islands, the vertebrate fauna is poor. It is remarkable for the absence of native mammals.

Functions of Phytogeography

Indian subcontinent lies between 8° and 37°N, and 68° and 97°E. It has its own peculiar physiographic, climatic and biotic features. It is surrounded in the South, east and west by the seas and in the north by the Himalayas. The subcontinent stretches between tropical and subtropical belts.

Climate: In the south and far-east the climate is typically tropical, while it is temperate in the north, and highly arid in the north-west. The climate of India is of monsoon type. Three distinct seasons occur in India:

1. *Cold Weather:* Season from December to February. During this time the mean temperature falls considerably low. In the north, snow fall occurs for days together and the temperature remains below 0°C for many weeks. In the south, temperature remains about 20°C. Some rainfall also occurs.

2. *Summer Weather:* The cold weather is followed by the hot, dry summer from March to June, as the temperature gradually rises, and the relative humidity decreases. The change of season is followed by leaf fall and blooming of trees particularly in Northern India. The temperature reaches upto 45°C or more in the arid

western region. Storms with dust and hot air are common during this season.

3. *Monsoon:* Towards the end of the summer by late June, the monsoon winds bring rain and the temperature gradually falls down. The rains start a little earlier in the eastern and southern parts of India.

The monsoon winds coming from the Bay of Bengal upto the Gangetic-Plains of the north result in 200 cm. of rainfall in the east and upto 100 cm in the west. By the end of September, the monsoon starts retreating, the sky becomes clear and the temperature gradually falls down, coming low to very low in December and January.

The variability in the temperature and rainfall patterns produces a great diversity in the vegetation of the subcontinent.

Zoogeography being an Importance Factor

Animal geography or zoogeography is concerned with the distribution of all the animals, invertebrates and vertebrates, the terrestrial and aquatic over the whole world. There are nearly 10,00,000 species of animals. It is practically impossible to cover up and study the distribution of each and every species. Most commonly the geography of land and freshwater vertebrates is taken into account, which constitutes just about 2 per cent of the total animal strength.

Distribution of animals can be studied at three levels-geographical distribution over the whole world, regional distribution in selected segments of the world and local distribution which includes geographical distribution of species in relation to each other and to ecology and evolution. But the zoogeography covers the distribution of animals over the whole world.

The area of distribution of animals is represented on map. Such a map shows the range of the family in question, and is called a distribution map. The map can be plotted depicting the distribution of a particular group or species or it represents the distribution of all the animals. The former are called simple maps and the latter as compound maps.

The distribution of animals varies greatly, but when range of one particular family unit is mapped the simple pattern that emerges falls into two main categories, *i.e.* continuous and discontinuous. Rats and cats among mammals, and hawks and cuckoos amongst birds have a continuous distribution and a wide range, *i.e.* they enjoy world-wide distribution. Crow family is another example of world-wide distribution but is comparatively less widely distributed, being absent from New Zealand. There are families which are continuously distributed but are confined to one continent or one part of the continent only (*i.e.* not found on several continents). For example, Giraffe family is confined to Australia and Marmoset Monkeys to South America.

The map depicts the discontinuous distribution of Tapir family, which is found in Malayan area and in South America. The two populations are widely separated by the Pacific Ocean.

There could be two possible methods of studying the geographical distribution of animals in the world:

1. Distribution of individual species over the world.
2. To divide the world into faunal regions and to describe their regional vertebrate fauna and their inter-relationship.

It is not conveniently possible to study the distribution of each and every species of family. Moreover, it will be beyond the scope of this book to enter into so much of details of distribution. Therefore, the second or alternative method has been followed. Depending upon the likeness of the fauna, the world has been separated into six main faunal regions. These are known as zoogeographical realms. These are:

1. Palaearctic
2. Ethiopian
3. Oriental

4. Australian
5. Nearctic
6. Neotropical.

The division was proposed by SCALTER for birds in a paper read before the Linnean Society in 1867. SCALTER'S regions were later on confirmed by WALLACE in 1876. The division is based upon the distribution of terrestrial and freshwater vertebrate fauna.

Many other regional classifications have been made, based either on the irregularities of distribution of one particular class of animals, or upon the distribution of temperature variations and similar climatic factors. None of these has been found so satisfactory as the above division of earth crust into regions by SCALTER and WALLACE. These combine reality as well as usefulness and hence are generally accepted.

Some zoogeographers suggest that Neotropical and Australian regions are zoologically different from the rest of the world and from one another that these two are equivalent of the remaining four put together. Thus according to this classification the earth crust is divided into three realms, Neogea (Neotropical), Notogea (Australian) and Arotogea (the rest of the world).

Another classification suggested by HEILPHIN in 1887, includes the amalgamation of Palaearctic and Nearctic regions and called Holarctic. The present discussion is based upon WAELACE'S concept.

Palaearctic Region and Importance

Geographic Limits: It is the northern part of old world. It extends over whole of Europe, China, Japan, Africa, North Sahara, Siberia, Mediterranean and Manchuria, Asia North of Himalayas and the North of Arabia. It is bounded by sea to the west, north and east and by Sahara and Himalayas to the South. The Palaearctic Region has, therefore, a continuous land connection with two of its neighbouring regions, the

Ethiopian and the Oriental. It is the biggest realm but without distinct boundaries.

Climate: Climate is chiefly temperate with an arctic fringe. It includes both wet forest lands and dry open Steppe land, large areas of coniferous forests and a fringe of Tundra. Eastern Asia including China and Japan contains deciduous forests. But most of the interior of Asia and North Africa are arid and open. Thus there is wide range of temperature variation and great fluctuation in rain fall.

Fauna: The fauna exhibits great variations in the climatic and vegetational subdivisions of the region. The fauna is richest in warmer areas and diminishes northward, until in the arctic area only the freshwater fishes, one frog (Rana), no reptiles and a few land and more freshwater birds and a few mammals are left.

The Palaearctic fauna in general exhibits affinities with Nearctic fauna and a number of genera and families are common to both regions. This suggests the existence of land bridges between Palaearctic and Nearctic regions in the ancient time. In all, Palaearctic region possesses 135 families of terrestrial vertebrates comprising 33 families of mammals, 68 of birds, 24 of reptiles and 10 of amphibians. Besides these, there are 13 families of freshwater fishes.

Fishes: The freshwater Palearctic fishes include several species of Cyprinids, a few localised Catfishes, a few Anabantids, Cobitids, Channids, some Percids and Mastacembelids. Five families represented by Perches, Sticklebacks, Sturgeons, Pikes and Toothed Sturgeons are similar to Nearctic region while Paddlefish occurs only in China and Dallia in eastern Siberia.

Amphibians: Amphibian fauna is rich and contains a large number of tailed amphibians. These are Necturus, Siren, Amphiuma and Proteus. The giant Salamander from eastern Asia is 5-1/2 feet long. Infact most of the tailed amphibians are found in Palaearctic and Nearctic Regions. The tailless amphibians are Frog and Toads including genera Rana, Hyla, Bufo and Rhacophorus.

Reptiles: The reptilian fauna is not very rich and there is not even a single genera, which could be described as endemic or exclusive. Even those species which are found in other regions are absent from this region.

There are a few lizards, skinks, tortoises and a few species of snakes. True vipers and Asian pit viper and colubrids are the only snakes found in Palaearctic Region. But in the southern edge of Palaearctic Region are found few other species. There are Testudo, Trionyx of eastern Asia, Alligator of China, Chameleon, Varanus, Typhlops and Leptotyphlops and Sand Boas are restricted to southern belt.

Birds: Palaearctic birds are represented by fifty three (53) families, out of which 17 occur more or less widely distributed (*i.e.* found in other regions also), while rest are migratory. These are Grebes, Loons, Rails, Hawks, Herons, Storks, Ducks, Wrenks, Cuckoos, Kingfishers, Swifts, Woodpeckers, Larks, Swallows, Thrushes, Jayas, Crows, Finches, Starling, Warblers and Old World flycatchers.

The Parrots are absent. Hedge sparrow are exclusively Palaearctic and Loons, Creepers, Grouse and Wax-wing are holarctic. This means that bird fauna of Palaearctic Region is in common with the neighbouring regions.

Mammals: Mammals in Palaearctic Region are represented by 33 families and numerous families of bats. But only two families of mammals (Myomorph rodents) are endemic. The important mammalian fauna includes Beavers, Moles, Shrews, Hedgehogs, Pandas, Pigs, Squirrels, Rabbit, Mice, Mustelids, Deer, Beavers, Dog and Cat families. Out of a total 33 families of mammals about one-third have a wide range of distribution, four families are shared with Nearctic and another four with the Ethiopian Region. The family of camel has a discontinuous distribution. The endemic families are Spalacidae (Spalax-mole rat) and Seleviniidae (Selevinia).

This suggests that vertebrate fauna of Palaearctic Region is not very rich and presents an overlapping of Nearctic fauna of New World and the Old World fauna.

The Palaearctic Region has been Subdivided into Four Subregions: European, Mediterranean, Siberian and Manchurian.

1. *European subregion:* It comprises of northern and central Europe, Black sea and Caucasus. It includes 85 families of terrestrial vertebrates. Amphibians and reptiles are represented by six families each. Only one genus of mammals, Myogale, is endemic to this region. Tits, Thrushes, Wagtails among birds and Wolf, Hedgehog, Shrews and Moles among mammals are very common.
2. *Mediterranean Subregion:* It includes remaining part of Europe, all the African and Arabian portions, Asia Minor, Persia, Afghanistan and Baluchistan. This region is supposed to be the richest part of Palaearctic Region, comprising of about 124 families of vertebrates. Upupa and Pastor among the birds and Elephant shrew, Civet, Hyaena, Hyrax and Porcupine among mammals constitute the characteristic fauna of subregion.
3. *Siberian Subregion:* It is represented by Northern Asia north to Himalayas and is characterised by unsettled and extremes of climatic conditions. Even then vertebrate fauna comprises of 94 families, of which families of Yak, Muskdeer and Mole are almost exclusively confined to this subregion. Phocaibirica (fresh water seal found in Baikal lake) is another characteristic form of this subregion.
4. *Manchurian Subregion:* It embraces Mangolia, Japan, Korea and Manchuria. Its fauna is rich and varied with 102 families and a number of peculiar forms. Among mammals Tibetan langur (Rhinopithecus), great Panda (Ailuropus), Chinese Water-deer (Hydropotes) and Tufted deer (Elaphodus) are represented in this region only.

Significance of Nearctic Region

Geographic Limits: It includes North America above tropics, Greenland, Newfoundland and Mexican Plateau. Except for a narrow strip of Central America it is completely cut off from all other regions by sea.

Climate: Nearctic Region resembles Palaearctic Region in its climatic conditions. As it extends from beyond the Arctic circle to the Southern tropic, it exhibits extremes of temperature and varied climatic conditions. It has extensive mountain ranges in the west running from North to South. In the north is the arctic belt of Greenland with layers of ice of unknown thickness. This is followed by coniferous belt, deciduous or mixed forests in the eastern part of North America, extensive grasslands (Prairies) in the central part and an arid zone in the south-west parts of Northern America.

Fauna: The Nearctic fauna as a whole, like Palaearctic one is much less rich than the fauna of other regions. It is mainly transitional, representing a mixture of fauna of Palaearctic and Neotropical Regions. Like Palaearctic, comparatively few families are found in this region and most of them are those which have a wide range of distribution. As recorded, there are about 120 families of vertebrates, of which about 26 families are of mammals, 4 of birds, 21 of reptiles and 14 of amphibians. Besides there are 24 families of fishes.

Fishes: Paddle fish (Polydon), 13 genera of Cyprinidae, some Catfishes, Garpike (Lepidosteus), Bowfin (Amia), Sturgeons, etc. are found in Nearctic Region. The Perches in this region are represented by the genera of Paralabrax, Huro, Pileoma, Bolesoma, Brytlus and Promotis; silurids by Hypodelus and Noturus, salmonids by Thaleichthys, etc.

Amphibians: Tailed amphibians dominate the amphibian fauna. Some of them are confined to this region only whereas others are shared with Palearctic Region. These are Amphiuma, Ambystoma, Axolotl, Siren and several other neotenous forms. About 9 genera of Salamandrididae are also resent. The tailless amphibians are Bufo, Hyla, Rana. The peculiar frogs and toads are Scaphiopus, Pseudacris and Acris.

Reptiles: Reptilian fauna is quite rich and presents a mixture of Palaearctic and tropical American groups. The common reptiles of Nearctic Region are—Trionyx, Musk turtle, Emydine, Alligator and Crocodile. Lizards are represented by Geckos, Ophisaurus, Phrynosoma and are found in Texas only. Helodrema is restricted to south-west Central America. Some other lizards are Uta, Uma, Euphryne, etc. Snakes are represented by Pit-vipers, Coral snakes and Colubrids. Specific snakes of the region are Rattle snake (Crotalus), Crotalophorus and those belonging to family Colubridae are Pituophis, Couophis and Chiloineniscus, etc. Four families of reptiles, *i.e.* snapping turtles, musk turtles, Phrynosorna and Helodernna are endemic to Nearctic Region.

Birds: The avian fauna is still less differentiated and presents a mixture of Palaearctic and Neotropical Regions. There are about 49 families of which 39 are widely distributed and rest may represent winter migrants from Neotropical or Palaearctic Regions. Other common birds of Nearctic Region are Grebes, Loons, Pelicans, Hawks, Herons, Vulture, Ducks, Quails, Rails, Cranes, Cuckoos, Plowers, Larks, Sandpipers, Gulls, Pigeons, Owls, Goat-suckers, Kingfishers, Swifts, Humming birds, Woodpeckers, Flycatchers, Wrens, Mockingbirds, Wax-wings and Creepers. The peculiar genera of this region are Chamaea, Catherpes, Centronyx, Picicorvus, Hylatomus, Philohela, Creagrus, Cupidonia, Ectopistes and Auriparus, etc.

Mammals: There are about 24 families of mammals and most of them are widely distributed. Primates and Monotermes are not represented here. Of the marsupials, only Virginian oppossum (Didelphys virginiana) is found in North America. Other commonly found mammals are Shrews, Rabbits, Squirrels, Moles, Pikas, Beavers, Cats, Bats, Bears, Deer and Bovids. Weasels, flying Squirrels, Jumping mice, Opossum, Armadillo, Tree Porcupine and Peccary are the immigrants from Neotropical Region. The exclusive forms of mammals found in Nearctic Region are American badger (Taxidea), Star-nosed more (Condytura cristata), Canadian porcupine, Long-legged bats and Leaf-nosed bats. WALLACE has divided Nearctid Region into four subregions: 1. California (western) Subregion; 2. Rocky Mountain Subregion; 3. Alleghany Subregion; 4. Canadian Subregion.

1. *California Subregion:* It embraces a narrow Strip of North America between Sierra, Nevada and Cascade ranges extending from Vancouver island and part of British Columbia. The fauna is very poor with 86 families of terrestrial vertebrates of which 21 are of mammals, 49 of birds, 8 of reptiles and 8 of amphibians. Three families are peculiar to the region, namely Haplodonfidae. Chamaeidae and Aniellidae. Vampire and free tailed Bats are characteristic of this region.
2. *Rocky Mountain Subregion:* It lies immediately to the east of California and embraces the dry and elevated area covered by mountains. It is the richest portion of Nearctic Region and contains 107 families of terrestrial vertebrates, but peculiar or endemic forms are absent. The characteristic animals of this region are Prong buck (Antilocapra), Rocky mountain goat (Haplocerus), American Bison, Prairie dog (Cynomys) and Poisonous beaded lizard (Heloderma).
3. *Alleghany Subregion:* It includes Eastern United States, *i.e.* east of Rocky Mountain Subregion and south of Great Lakes. This Subregion is characterised by the following peculiar animals: Opossum, Star-nosed moles and Vampire bats among mammals; Passenger pigeons and Carolina parrots among birds. Turkeys (birds) and Sirenids (mud eels) are the other two peculiar families.
4. *Canadian Subregion:* It comprises of remaining portion of North America and Greenland. It is known for its poor fauna but it exhibits great resemblance to Palaearctic fauna. The peculiar forms are Reindeer, Bison, Sheep, Lemmings, Gluttons, Polar bears, Elk and Arctic fox.

4

ACTIVE INSECTS

Lac insects and their products have been known to naturalists since very early times. The lac has been referred in ancient Sanskrit works, *viz.*, Atharva-Veda (Dave, 1950; Hora, 1952) and was called as 'Luxa'. It is mentioned in Mahabharata that 'Luxa Griha' was made up of lac which was prepared by Kaurava for Pandavas. Abul Fazal (1590) in his famous book 'Ain-i-Akbari' has mentioned in detail about the lac industry in India.

Mahdihassan (1950, 1952), has referred about the lac insect and its products in China. The first scientific reference regarding the lac and lac insect is the report of Kerr and Glover in 1782. Subsequently, much work has been done by various workers on the organisation, distribution, taxonomy, host plants, culture, production, enemies, chemistry and technology.

Three products from lac insects, *viz.*, the lac-dye, lac-wax and lac (resin) have been items of trade and commerce.

Distribution: India has its monopoly on the production of lac. Other countries like Africa, Australia, Brazil, Burma, Sri Lanka, China, Formosa, France, W. Germany, Japan, Malaya,

Nepal, Spain, Thailand, Turkey, U.S.A. and some others also produce lac. But in Thailand, Malaya, Burma and Nepal the lac producing industries are increasing day-by-day. Thailand has become the main competitor of India in export of lac.

In India major lac producing places are Assam (Kashi Hills), Bengal (Calcutta, Jangipur, Murshidabad, Mathrapur, Malda), Jharkhand (Manbhum, Palamau, Ranchi, Santhal Pragana), Delhi, Gujarat, Hyderabad, Kashmir, Madhya Pradesh (Rewa, Umaria), Chattisgarh (Damoh, Champa, Bilaspur), Chennai, Coimbatore, Mysore, Orissa (Cuttak, Mayurbhanj), Punjab (Hoshiarpur, Shahpur), Rajasthan (Indergarh, Kota, Jaipur, Jhallawar, Karauli), and Uttar Pradesh (Ghazipur, Mirzapur, Agra), etc.

Lac Insect

Phylum .... Arthropoda

Class .... Insecta

Order .... Hemiptera

Sub-order .... Homoptera

Super-family .... Coccidae

Family .... Leciferidae

Genus .... *Tachardia*

Species .... *lacca*

Lac insect (*Tachardia lacca)* previously known as *Laccifer lacca* is a minute, resinous, crawling scale-insect which inserts its beak into plant tissues, sucks juices and grows, and secretes lac from the hind end of the body. Its own body ultimately gets covered with lac in the 'CELL'. Lac is actually secreted for its protection and not for the food of the insect. The commercial lac is produced in large quantities by female as a protective covering of its body which is injurious to the host plants.

Male: Male is red in colour and 1.2 to 1.5 mm in length. It secretes bright creamy lac. It has reduced eyes and ten segmented antennae. The mouth-parts are of piercing and sucking type. Thorax bears three pairs of legs and one pair of hyaline wings. The abdomen is eight segmented and terminates into a short, chitinous prominent general sheath containing penis. On either side of this genital sheath a white elongated caudal seta is found.

Female: Female is larger than males and measures about 4 to 5 mm in length. The pyriform body of the female is enclosed in a resinous cell. The head, thorax and abdomen are not clearly distinct. The mouth-parts are of piercing and sucking type. The antennae are clearly visible and degenerated. The posterior end of the body has a median and two lateral processes. The legs are in degenerated·form.

Pest Control

The Entomology consist from "Greek" words. The means of "Entomo" is insect and means of "Logos" is knowledge.

- Entomology is the branch of zoology that deals with the study of insects under class insecta of phylum "Arthropods".
- The lac insect *Laccifer lacca* Kerr belongs to the order Hemiptera.
- The silk worms belong to the insect order Lepidoptera.
- Honeybees belong to the order of "Hymenoptera".
- It has been estimated that 50-70% of the grain crops are pollinated by insects, a major portion of it by bees alone.
- The latest production of honey in India is estimated at about 25000-30000 tonnes, its major portion obtained from "Apis *dorsata*".
- The all insects deficit RBC in blood.
- The blood colour of insects is green with yellow.
- The latest estimated total annual losses caused by insect pests to major crops and the food-grains in storage in India to the Rs. 336.6 billion.

- The world wide pre-harvest and post-harvest food losses due to pests have been estimated to be about 48%.
- There is a higher percentage of 46% loss for rice and a lower percentage of 24% for wheat.
- The higher loses occur in Africa 42% and Asia 43% respectively of their potential crop value.
- By the controlling insect-pests of sugarcane the yield could be increased by 20% from 229.2 to 284.9 million tonnes leading to an additional income of 43730 millions of rupees.
- Plant genetic system, a Belgium Biotechnology company in July 1987 were the first reported the development of transgenic plants of tobacco.
- Over 2.5 billion kg of pesticides are applied in the world annually.

Table: Pesticides Consumption durint 1999-2000 in India

Insecticides	60%
Fungicides	21%
Herbicides	14%
Others	5%

- Punjab have the major consumption technical grade pesticide in India, approximately 7150 metric tonnes.
- Lowest consumption of pesticide in India is Daman and Diu, approximately 1 metric tonnes.
- Bartlett (1956) coined the term "Integrated Pest Control"

The term "pest management" was given by Geier (1970).

Control of Integrated Pest: IPM as a dynamic and constantly evolving approach to crop protection in which all the suitable management tactics and available surveillance and forecasting information are utilised to develop a holistic management programme as part of a sustainable crop production technology (By Dhaliwal and Arora 2001)

- Destructive Insect and Pest Act was enforced in the year of 1971.
- The maximum percentage of insecticide is used in cotton in India.
- Average damage of food grains during stored period in India is 10%.
- Plant Protection and Quarantine Act was passed in the year of 1914.
- First insecticide discovered was D.D.T.
- The maximum amount of sugar in honey is 73%.
- 68% resin found in lac insect. *(U.P. Lecturship 2003).*
- Central Honey Bee Research Station is situated at Pune (Maharashtra).
- Lac Research Station is situated at (Bihar).
- Central Silk Research Station is situated at West Bengal.
- D.D.T. was first of all made by "Zedler" scientist.
- The first synthetic pyrethroid was described in 1973.
- LD_{50}: Lethal dose required for killing 50% population of test animals is known as LD_{50} value. The unit of LD_{50} is mg/kg of body weight of test animals or Hg/g body weight of insect.
- Biology has two subdivisions. BOTANY-which deals with plants and ZOOLOGY-which deals with animals.
- The aim of zoology is to collect knowledge of animals.

Helping Plants

The lac insects have more than one host plant. The selection of suitable host plant for the cultivation of lac is of much importance. To establish the lac industry one should know well about the topographic and climatic conditions for the growth of host plants suitable for that particular region. Brun (1958) has mentioned that 113 varieties of host plants are found in the geographical Indian regions including Pakistan and Burma. Out of these 113 host plants only 14 are very common in India which are as follows:

1. Kusum — *Schleichera oleosa*
2. Babul — *Acacia nilotica*
3. Ber — *Zizyphus mauritianas*
4. Palas — *Butea monosperma*
5. Ghont — *Zizyphus xylopyra*
6. Khair — *Acacia catechu*
7. Peepal — *Ficus religiosa*
8. Gular — *F. glomerata*
9. Pakapi — *F. virens*
10. Putkal — *F. globella*
11. Mango — *Mangifera indica*
12. Sal — *Shorea robusta*
13. Shisham — *Dalbergia sisso*
14. Fig — *Ficus carica*

The quality of lac is directly related with the quality of host plant. So far, no artificial product has been able to replace the lac. Khair, Kusum and Babul give better quality of lac when sown directly in the field. But Palas, Ber and Ghont give good crop when they are first sown in nursery and then transplanted to the lac growing field. Palas and Ber produce a particular type of lac which is called as 'KUSUMI LAC'.

Cultivation of Lac: Lac cultivation is a complicated process, so the cultivators should know well about the inoculation, swarming period and harvesting of lac.

Inoculation: The first procedure in the lac cultivation is the inoculation of lac insect. Inoculation is the process by which young ones get associated properly with the host plant. Inoculation is of two types:

Natural Inoculation: The inoculation taking place in normal routine or in natural way is very simple and common process during which the swarmed nymphs infect the same host plant again and start to suck the juices from the twigs. The natural incubation of swarmed nymphs has some drawbacks which are as follows:

(a) *Incomplete Nutrition:* Lac insects with their piercing and sucking mouth-parts, pierce into succulent twigs and suck the cell sap of the same host plant for nutrition. If the cell sap of the same host plant is further sucked out by the swarmed nymphs of the second crop continuously, the growth of the host plant would be retarded. In this way lac insect may not be able to get enough nutrients from the same host plant. The lac insects due to lack of sufficient nutrients lose their proper development, thereby affecting the production of lac also.

(b) *Irregular Inoculation:* During the natural inoculation it is not sure that uniform sequence of inoculation takes place. If inoculation is not of continuous fashion, a regular crop of lac may not be obtained.

(c) *Unfavourable Climatic Conditions:* At the time of swarming a number of factors like high intensity of sunlight, heavy rainfall, flow of wind, etc. affect the proper inoculation of nymphs. These natural environmental factors may also affect the host plant at the same time and may cause a gap of inoculation resulting in irregularity of the lac crop.

(d) *Multiplication of Parasites and Predators:* Lac insects have certain enemies in the form of parasites and predators. If the crop is not harvested in time and lac is allowed to remain on the same twig, the multiplication of parasites and predators takes place which hampers the population growth of lac insects.

Thus, keeping in view the above drawbacks the natural procedure of inoculation is avoided and certain devices have been developed to ensure artificial method of inoculation.

Artificial Inoculation: The main idea behind the artificial method of inoculation is to check all possible drawbacks of natural inoculation.

In this method first of all host plant should be pruned in January or June. The twigs bearing insect nymphs which are about to swarm, or just before swarming are cut in sizes ranging between 20 to 30 cm in length. Then the cut pieces of these twigs are tied to fresh trees in such a way that each stick touches the tender branch of the tree at several places which form bridges for the migration of the nymphs. After swarming, these twigs should be removed and separated from the host plant. The following precautions should be taken in artificial inoculation:

(i) One must ensure that the twigs, which are going to be tied on fresh host plant, are having good number of nymphs or eggs. It is also possible that from many of the twigs nymphs have swarmed out, thus inoculation would prove unsuccessful.

(ii) The twigs provided with eggs or nymphs should be without any parasite and predator.

(iii) The eggs or nymphs present on the twigs should be healthy and about to swarm so that one has not to wait for longer period and thus save time.

(iv) For the uniformity of inoculation, 3 to 4 twigs should be utilised.

(v) Host plants should be changed from time to time for the proper nutrition of the nymphs.

These insects are very small and if they move to a long distance there are chances of mortality of the nymphs. Due to maximum contact of twigs, swarming nymphs have not to move for long distance and find suitable places to establish on the host plant.

Inoculation Period: In India two types of crops, *viz.*, Rangini and Kusumi are grown in a year. The Rangini crop is of two types called as Kartiki and Baisakhi crop which produce Kartiki and Baisakhi lac respectively. The Kusumi crop is also of two types, *viz.*, Agahani and Jethi which produce Agahani and Jethi lac respectively.

Thus, the inoculation periods of all the four types of crops are different. The inoculations of Kartiki, Baisakhi, Agahani and Jethi crops are recommended in months of June to July, October to November, July and January to February respectively. But if continuously four crops are taken, the plant would not get any rest which may cause less production of lac.

Swarming: It is very important phase in the life history of lac insect. So one should have accurate knowledge about the actual date of the swarming. At the time of swarming, the upper surface has yellow spot on the anal region. At this stage muscle contracts and insect gets detached from the place of attachment. Thus, it leaves a hollow cavity which later on gets covered with wax also. When these eggs are to be hatched out they become orange coloured. Thus, it is an indication that swarming has taken place. Thus, by trials and learning methods, *i.e.*, by practice one could know about the exact date of swarming by looking at the colour of the eggs.

Harvesting of Lac: The process of collection of ready lac from host tree is known as harvesting. In common practice the harvesting is of two types:

1. *Immature Harvesting:* The harvesting of the lac before swarming is called as immature type of harvesting and the lac thus obtained is known as 'ARI LAC'.
2. *Mature Harvesting:* The collection of crop after the swarming is called as mature harvesting and the lac obtained is known as 'MATURE LAC'.

The harvesting of lac before the swarming has some drawbacks because the lac insects may be damaged at the time of harvesting which would affect the population of lac insects and ultimately result in great economic loss to the cultivators. But in case of palas lac (Rangini lac) it is found that Ari lac gives better production. Therefore, Ari lac harvesting is recommended in case of palas only. In all other cases immature harvesting should be discouraged. It is also found that in cold areas mature crop yields better quality of lac.

Harvesting Period: The harvesting periods of different crops are quite different in accordance with the inoculation of crops. Kartiki crop is harvested in October to November whereas, Baisakhi crop in May and June. The other crops like Agahani and Jethi are harvested in January to February and June to July respectively.

Recent Plan for Lac Cultivation: With the increasing number of lac industries some advanced plans have been recommended for the better cultivation of lac crops. Two types of planning are used nowadays.

1. *Ceupe System:* All the trees of host plants of a definite area are not used under continuous cultivations process of lac crop because if all host plants of a farm would be under continuous attack of lac insects, 100% plants may not get any rest and thus the production of the lac would be affected due to deficiency of nutritive cell sap to the swarmed nymphs and adults. So, the plants of a farm are numbered into 5 groups of plants. This artificial division or marking of trees is called as ceupe system of crop cultivation. In this system when one group of host plants is under the process of cultivation of lac, other groups of host plants would be under rest.
2. *Alternation of Plant:* In this system the variety of host plant is changed after one crop. So, swarmed nymphs are inoculated on the tree of other variety of host plant. In this way every host plant can get enough rest resulting into better production of lac.

Processing of the Lac Industry: When the crop matures fully most of the lac is harvested and some part is left on the host plant. For the proper cultivation, the host plant should be pruned in January every year.

The twig bearing the lac alongwith eggs is called a BROOD LAC STICK and lac is known as BROOD or STICK LAC. The processing starts with the scraping of the stick lac from the twig. The scraped lac is subjected to removal of many impurities like dead parts of the lac insects, eggs and colouring matter, and finally crushed by hand-operated mortars.

Then the material is air dried and obtained in the form of granules which is known as SEED LAC. This seed lac is soaked in water, washed, dried in sun light, bleached and heated to melt on charcoal fire in cloth bag of 3 to 4 metre. At the time of heating the bag is twisted and the lac is squeezed out of the bag. The impurities of the lac are left out in the bag, and are called as KIRRI LAC.

The squeezed lac is now allowed to cool and solidify around the button-shaped forms which is now called BUTTON LAC or PURE LAC. This pure lac when stretched into thin sheet is called as SHEET LAC. This sheet lac when dissolved in water, produces white or orange coloured lac which is called as SHELL LAC. Shell lac is, in fact, prepared by boiling the seed lac with yellow arsenic in a certain proportion. Thus the shell lac is most purified form of lac.

The quality of lac depends upon the host plant. Kusumi lac is said to be the best lac while Dhak is supposed to be the worst and cheapest one. The quality and colour of the lac is variable according to the presence of gum and resins in the host plants.

Composition of Lac: Lac is a complex substance having large amount of resins, together with sugar, water and other alkaline substances. The percentage of various constituents are as given below:

(1) Resin — 68 to 90%

(2) Dye — 2 to 10%

(3) Wax — 6%

(4) Albuminous matter — 5 to 10%
(5) Mineral matter — 3 to 7% and
(6) Water — 3%

Properties of Lac

(1) Lac is not soluble in water but easily soluble in alcohol. This property of lac has great value for insulation of electrical connections.
(2) Lac is easily fusible on heating.
(3) Lac has adhesive quality.
(4) It has binding property when mixed with alcohol.
(5) Lac is also soluble in weak alkali like ammonia.
(6) Lac is a bad conductor of heat.

Enemies of Lac Cultivation: Lac cultivation is destroyed by biotic and abiotic factors:

1. *Abiotic Enemies:* These are high intensity of light, high temperature, high humidity, heavy rainfall and flow of wind.
2. *Biotic Enemies:* The main biotic enemies of lac cultivation are mammals and insects. Krishnaswami *et. al.* (1957, 59), and Gepulpure et. al. (1963), have reported that squirrel, rats, and monkeys cause great damage to the lac crop.

The insects are very powerful enemies of lac crop. Annual loss due to the insect enemies is to the tune of about 4 lakh maunds. The insects damage the crops in different ways.

Parasites: The lac insects are parasitised by eight species of chalcidoid parasites like, *Parenchthrodryinus clavicornis, Erencyrtus dewitzii, Tachardiaephagus tachardiae,Tachardiaephagus tachardiae* var. *somervilli, Eupelmus tachardiae, Coccophagustschirchii, Mariettajavensis* and *Tetrastichus purpureus.* These parasites lay their eggs into lac insects and parasitised 4.8 to 9.9% of lac insects per year and 1/3 of the parasitised cells are males. Thus, it may be concluded that parasitisation is not a major cause of the damage to the lac cultivation.

Predators: Predators cause very severe damage to lac cultivation and 35% of the lac cells are damaged by two predators, *viz., Eublemma amabilis* Moore (Lepidoptera: Noctuidae) and *Holocera pulverea* Meyr (Lepidoptera: Blastobasidae). Female lays eggs near encrusted twigs from where larva emerges and feeds on lac insects.

THE CAREFULNESS

(1) Twigs for inoculation should be cut just before the swarming to get healthy brood.

(2) Twigs used for inoculation should be free from predators and parasites.

(3) Twigs tied for inoculation should be removed from inoculated host plants after a maximum period of 20 days.

(4) Lac left on the host tree for swarming should be removed in October and November.

(5) The brood lac after swarming should be destroyed along with predators and parasites on it.

(6) The lac scraped from the tree should be taken away from the area of lac infected trees.

(7) Fumigation and water immersion of lac, before removing from twig, should be done.

Lac Industry: India used to produce about 97 per cent of the total lac output in the world but at present it has come down to 50-60 per cent. The cultivation of lac has been good source as an earner of foreign currency. About 50 per cent of the total lac produced in India is obtained from Chhotanagpur area. States like Orissa, Punjab, Madhya Pradesh, W. Bengal, Uttar Pradesh, Gujarat, Rajasthan, Assam, etc. are increasing the production of lac Nowadays. On a very small level lac producing is also reported from Delhi and Kashmir. The average yearly yield of lac in India is about 15,000 metric tons. A lac research institute 'Indian Lac Research Institute' Namkum, Ranchi had been established in 1925 which is producing good quality of white lac. The Indian white lac is supposed to be better than red or other coloured lac because they produce stain or spots at places where they are kept.

This is mostly small scale industry with around 350 factories, mostly located in Bihar. In Mirzapur district alone there are about 40 factories. Out of total lac produced in India about 85 to 95 per cent is exported specially to Britain, U.S.A., Russia and West Germany.

Practical Appearances

In 19th century lac dye was in more use than lac resin. Presently due to availability of a better and cheaper annaline dyes the use of lac as a dye has been discarded. The manifest uses of lac is one of the Nature's standing gifts. The various used to which it is put are:

(1) It is utilised in the preparation of gramophone records. Previously this industry utilised major part of the lac produced annually. But nowadays to a great extent plastic is being used in this trade.

(2) It is of utility to Jewellers and Goldsmiths who use lac a filling material in the hollows in gold ornaments like bracelets, armlets and necklaces, etc.

(3) It is an essential gradient used extensively for making polishes, paints and varnishes for finishing wooden as well as metal furnitures and doors, etc.

(4) It is utilised for the preparation of toys, buttons, in pottery and artificial leather.

(5) It is used in the manufacture of photographic material, lithographic ink and for stiffening felt and hat materials.

(6) It is used as an insulating material for electrical goods.

(7) It is also used in confectionery trade as antifowling for applying on ship bottom, grinding stone industry and for ammunition and fire works.

(8) Last but not the least used commonly as sealing wax.

Thus, it is of great use and considered to be as one of the cash crops for the cultivators and also to the Government as source of foreign exchange earners which amount to crores of rupees annually.

Classification of Zoology

Entomology	—	Study of Insects
Protozoology	—	Study of Protozoa
Nematology	—	Study of Nematodes
Rodentology	—	Study of Rodents
Ornithology	—	Study of Birds
Mammology	—	Study of Mammals
Ichthyology	—	Study of Fishes
Helminthology	—	Study of Worms
Acarology	—	Study of Mites

Classification of Entomology

Insect Taxonomy

Insect Morphology

Insect Anatomy

Insect Histology

Insect Physiology

Insect Toxicology

Insect Pathology Insect Parasitology

Insect Genetics

Entomology in Use

Agricultural Entomology

Medical Entomology

Horticultural Entomology

Veterinary Entomology

Storage Entomology
Forest Entomology
Industrial Entomology

Structural Pattern of Insects

The Rasping-sucking mouth parts are intermediate in structure between the piercing-sucking and chewing type and are found in Thrips. The Chewing-lapping type of mouthparts are found in honeybees. The Piercing-sucking type of mouthparts are found in bugs and mosquitoes. The Sponging-type of mouth parts are common in flies the typical example being that of the housefly.

- *Pest:* A pest is an organism whose population often increase above a certain level of economic injury and its existance conflicts with man's welfare, convenience and profits.

Sorts of Pests: There are three categories of pests

(i) *Regular Pests:* These are generally found in abundance during a crop season e.g. aphids, Jassids, and Thrips.

(ii) *Sporadic Pests:* These assume pest status occasionally in certain years and include locusts, grasshoppers, hairy caterpillars, crickets and cutworms.

(iii) *Potential Pests:* These pests normally cause negligible damage but may become highly destructive resulting from some disturbance in the environment and the consequent increase in their number e.g., armyworm on wheat.

Step to Control Insect-Pests

(a) *Legislative Measures:* Legislative control involves enactment of laws to regulate the entry, establishment and spread of pests.

Legislation is also imperative to stop the accidental entry, from outside the country of certain pests which may not be present in that country.

Thus legislation is of four kinds:

(i) Legislation for foreign quarantine to prevent the introduction of new pests from abroad.

(ii) Legislation for domestic quarantine to prevent the spread of established pests within the country or within a particular state.

(iii) Legislation for notified campaign of control against pests.

(iv) Legislation to prevent the adulteration and mishandling of insecticides or other devices used for the control of pests.

(b) Cultural Practices: The cultural methods of insect control comprise regular form operations, which are so performed as to destroy the insects or to prevent them from causing injury.

For achieving cultural control, the various agricultural practices can be categorised under various groups

1. Tillage	2. Clean seed
3. Irrigation	4. Fertilizers
5. Clean culture	6. Crop rotation
7. Trap crops	8. Pruning and Thining
9. Intercropping	10. Destruction of crop residues.

(c) Mechanical and Physical Control: Mechanical and physical control measures involve the use of force or physical factors of the environment with or without the aid of special equipment.

The Functionaries

It means working with hands. Sometimes with the aid of some simple equipment, like bags, nets, etc.

(d) *Chemical Control:* The use of chemicals for the control of insect-pests. The classification of insecticides is as follows.

Classification of Insecticides

Insecticides can be grouped in various ways:

According to the Mode of their Entry

(1) *Stomach Poisons:* They are generally used against insects with chewing type of mouthparts and under certain conditions against those with sponging, siphoning, lapping or sucking mouth parts.

(2) *Systemic Poisons:* A systemic insecticides when applied to seeds, roots, stems or leaves of plants is absorbed and translocated to various parts of the plant in amounts lethal to insects which feed on them, like-dimethoate, phosphamidon, phorate, aldicarb, etc.

(3) *Contact Poisons:* The contact poisons are applied as sprays or dusts either directly onto the body of insects or to the places frequented by them. These poisons kill the insects either by clogging spiracles and respiratory system or by entering through the cuticle into the blood and acting as nerve or general tissue poisons. like-chlorinated hydrocarbons (DDT, HCH, aldrin, etc.) carbamates (carbaryl, aldicarb, carbofuran, etc.) and organophosphates (Malathion, Monocrotophos, etc.)

(4) *Fumigants:* Poisonous gases, derived from either solids or liquids are used as fumigants to kill insect pests of stored grains and other products in warehouses, museum, godowns, etc.

- The common fumigants are-sulphurdioxide, phosphine, methylbromide, EDB, etc.

According to Mode of Action

Physiological poisons

Protoplasmic poisons

Respiratory poisons

Nerve Poisons

According to their Chemical Composition

(i) *Elements:* Such as sulphur, phosphorus, thallium, mercury, etc.

(ii) *Inorganic Compounds:* Such as lead arsenate, sodium fluosilicate, zinc phosphide, paris green, etc.

(iii) *Organic Compounds:* (a) Compounds of plant origin such as pyrethrum, nicotine, rotenone, etc.; (b) Animal and mineral oils such as fish oil, diesel oil, etc.; (c) Synthetic organic compounds such as DDT, Malathion, carbaryl, etc.

(iv) *Poisonous Gases:* Such as hydrogen cyanide ethylene dichloride, carbon tetrachloride, methyl bromide, phosphine, etc. used as fumigants.

(a) *Nematicides:* These chemicals are used for controlling plant parasitic nematodes. Some of common nematicides are mentioned here.

(1) DBCP	(2) D-D Mixture
(3) Ethylene dibromide	(4) Ethoprop
(5) Phorate	(6) Aldicarb

(b) *Rodenticides:* These chemicals are used for control of rodents such as-strychnine, warfarin, zinc phosphide, etc.

(c) *Molluscicides:* Molluscisides may be classified into two broad categories.

(i) *Aquatic Molluscicides:* copper sulphate niclosamide, sodium pentachlorophenate, etc.

(ii) *Terrestrial Molluscicides:* include carbamates-aminocarb, methiocarb, mexacarbate, etc.

(d) *Fungicides:* These chemicals used against fungus like: (i) Bordeaux mixture; (ii) Sulphur and allied compounds; (iii) Carbamates; (iv) Organic mercury compounds; (v) Antibiotics; (vi) Nitrogen compounds Pesticide Application Equipment.

The various equipment used include dusters, sprayers,

vapour and smoke generators, agricultural aircrafts, granule applicators, etc.

The Insecticide Dusters: The insecticides dust which are very fine particles when falling free in air either slowly or driff for a long distance due to wind.

Parts of the Dusters

(i) *Hopper or Container:* It is used for holding the dust.

(ii) *Blower:* It work to throw the dust outside from container to the discharge outlet.

(iii) *Handle:* It is used to rotate the gear and there by fan throws the air from blower.

(iv) *Feeding Brush:* It agitates the dust in the hopper to prevent uneven dusting.

(v) *Feed Mechanism:* It control the rate of powder coming from hoppers.

(vi) *Discharge Line:* It carries the powder from bowler to nozzle.

(vii) *Shoulder Streps:* Shoulder streps are provided with mounting frame in some dusters to set whole mechanism on the chest.

Classification of Dusters

(a) Manually Operated Dusters
  (i) Plunger duster
  (ii) Bellows dusters
  (iii) Rotary or fan dusters
  (iv) Wet dusting equipment

(b) Power Operated Dusters
  (i) Tractor mounted duster
  (ii) Engine-operated duster

Merits of Dusters

1. Large area can be covered
2. Less expensive

3. Easy to operate
4. Very effective in arid zone area Demerits
5. Loss of chemicals due to drift problem
6. Less efficient
7. Increase toxic hazzard
8. Sprayers

A sprayer is an appliance which atomises the spray fluid, which may be a suspension, an emulsion or a solution, the fluid is ejected with some force for proper distribution. The various types of sprayers can be grouped into four categories, depending upon the volume of spray fluid discharge to cover a unit area.

1. *High Volume Sprayer:* These require 300-500 litres per ha. of spray fluid. For example-knapsack sprayer, stirrup sprayer, rocker sprayer, compression sprayer, foot sprayer, etc.
2. *Low Volume Sprayer:* These sprayers require 50-100 litres of spray fluid per ha. for instance-motorised knapsack sprayers.
3. *Ultra Low Volume:* These sprayers require 1-5 litres per ha. of spray fluid. For example knapsack mist blowen fogair.
4. *Aerosols:* Less than 1.0 litres per ha. of spray fluid is required to create fog. The example are-pressurised containers, swing fog machine.

Types of Sprayers

Manually Operated Sprayers

These are of two types:

(i) *Pneumatic sprayers:*
   (a) Hand compression sprayers
   (b) Pressure retaining pumps.
   (c) Small pneumatic atomisers.

(ii) *Hydraulic sprayers:*

(a) Knapsack sprayer (b) Foot sprayer

(c) Rocker sprayer (d) Stirrup sprayer

Power Operated Sprayers

(i) Mist blower or motorised knapsack sprayer
(ii) Tractor mounted sprayer
(iii) Power sprayer
(iv) Ultra-low-volume sprayer
(v) Aerosol dispensers
(vi) Smoke generator
(vii) Vapour generators.

Agricultural Aircrafts

(i) Light aircraft
(ii) Medium aircraft
(iii) Heavy aircraft

Granule Applicator

Placement of insecticidal granules in the whorls of crops like-maize, jawar, etc., is very effective for the control of stem borers.

Soil Injectors

Most soil fumigants are liquids and are applied with soil injectors.

Merits of sprayers

1. Less drift problems as compared to dusters.
2. It is handy and simple to use.
3. Used to apply various types of chemical such as bactricide, fungicide, insecticide, etc.
4. Most commonly used appliances.
5. Covers more area in less time.
6. Treatment can be done on height and also blow the leaves.

Sustainable pest management is otherwise known as IPM, i.e., Integration of tactics of control a single pest on one or more crops. IPM as applied in agriculture is ideally the use of most effective, economical, safest, ecologically sustainable and sociologically acceptable combination of physical, chemical and biological methods to limit the harmful effects of crop pests. According to FAO panel of experts (1966-72) IPM is-"A pest management system that in the context of the associated environment and the population dynamics of the pest species, utilises all suitable a manner as possible and maintains the pest population at levels below those causing economic injury." The overall objective of IPM is to create and to maintain situations in which insects are prevented from causing significant damage to crops.

A number of tools or components have been successfully used for IPM in several crops. These include-use of pest resistant or tolerant, predators and pathogens, use of parasites, summer ploughing, quarantine measures, late planting, hard collection and destruction, judicious use of pesticides, attractants, repellents, sterilants, growth regulators. The sustainable pest management in India has yet to come to be well recognised, but it is still more of an aspiration than a reality for the average farmers.

Biopesticides and their Significance: Biopesticides are most important, effective and commercially viable as these are inexpensive cause no pollution. Pose no risk to human health and have a long term capability to control pests. Biopesticide agents are abundantly available in nature. Several pathogens including viruses such as-NPV (Nuclear polyhedrosis viruses and granulosiss viruses (G.V.)

- Bacteria like-*Bacillus thuringiensis.*
- Fungi like-Metarhizium, Beauveria, Verticillium.
- Protozoa like-*Schizogregrine* cause diseases in insects and destroy them. Similarly, several insect-parasitoids (Parasites thriving on insects are also known in nature.
- *Trichoderma* is an egg parasitoid of several pests.

	Common Name	*Scientific Name*	*Order*	*Host Plants*	*Damaging Stage*	*Mouth Parts*
1.	Locust	*Schistocera gregaria* Forsk	Orthoptera	Polyphagous adult	Nymph and cutting type	Chewing and
2.	Termite	*Odontotermes obesus* Ramb	Isoptera	Polyphagous	Nymph and adult	Chewing and cutting type
3.	White Grub	*Holotrichia consanguinea*	Coleoptera	Polyphagous	Grub and adult	Biting and cutting type
4.	White Fly	*Alerolobus baradensis* Mask	Hemiptera Homoptera	Sugarcane, Jawar, Bajra, Wheat, Barley	Nymph and adult	Piercing and sucking type
5.	Leaf Hopper or Pyrilla	*Pyrilla perpusilla* Walk	Hemiptera Homoptera	-do-	-do-	-do-
6.	Sugarcane top borer	*Tryporyza nivella*	Lepidoptera	Sugarcane, Jawar, Bajra, Kans, Sarkanda	Larva	Biting and cutting type
7.	Shoot borer	*Chilo infuscatellus* Snell	Lipidoptera Jawar, Maize	Sugarcane,	Larva	Biting and cutting type

Contd....

	Common Name	***Scientific Name***	***Order***	***Host Plants***	***Damaging Stage***	***Mouth Parts***
8.	Paddy stem borer	*Tryporyza incertulas*	-do-	Paddy	-do-	-do-
9.	Rice gundhi bug	*leptocorisa varicornis*	Hemiptera	Paddy, Jawar, Bajra Maize	Nymph and adult	Piercing and sucking type
10.	Paddy plant hopper	*Nephotettix* spp.	Hemiptera	Paddy	-do-	-do-
11.	Cotton Jassid	*Amrasca bigutulla bigutulla*	Hemiptera	Cotton, bhindi	Nymph and adult	-do-
12.	Cotton leaf roller	*Sylepta derogata*	Lepidoptera	-do-	Larva	Chewing and cutting type
13.	Spotted boll worm	*Earias vitella*	-do-	-do-	-do-	-do-
14.	Pink boll worm	*Pectinophora gossypiella* Sound	Lepidoptera	-do-	-do-	-do-
15.	Mustard aphid	*Lipaphis erysimi*	Hemiptera	Radish, Rai, Mustard, Cauliflower, turnip, etc.	Nymph and adult	Piercing and sucking type

Contd....

Common Name	Scientific Name	Order	Host Plants	Damaging Stage	Mouth Parts
16. Painted bug	*Bagrada cruciferorum*	Hemiptera	-do-	-do-	-do-
17. Cut worm	*Agrotis ypsilon*	Lepidoptera	Mainly gram and potato	Larva	Cutting and biting type
18. Gram caterpillar	*Helicoverpa armigera*	-do- tomato, Arhar	Gram, pea,	-do-	-do-
19. Red pumpkin beetle	*Raphidopalpa foveicollis*	Coleoptera	Plants of cucurbitaceae	Grub and adult	-do-
20. Fruit fly	*Dacus cucurbitae*	Diptera	Plants of cucurbitaceae	Maggot	-do-
21. Epilachna beetle	*Epilachna vigintioctapun-ctata*	Coleoptera	Brinjal and cucurbits	Grub and adult	-do-
22. Brinjal shoot and fruit borer	*Leucinodes arbonatis*	Lepidoptera Potato	Brinjal and	Larva	-do-
23. Diamond back moth	*Plutella maculipensis*	Lepidoptera cruciferae	Plants of biting type	-do-	Cutting and

Contd....

	Common Name	Scientific Name	Order	Host Plants	Damaging Stage	Mouth Parts
24.	Tobacco caterpillar	*Prodenia litura*	Lepidoptera	Tobacco and cruciferae	-do-	-do-
25.	Mango hopper	leaf *Amritkodus atkinsoni*	Hemiptera Homoptera	Mango adult	Nymph and sucking type	Piercing and
26.	Mango mealy bug	*Drosicha mangifera green*	Hemiptera Homoptera	Mango, Jack fruit, etc.	-do-	-do-
27.	Lemon buterfl	*Popilio demoleus* Linn	Lepidoptera	Citrus	Larva	Biting and cutting type
28.	Ber fruit fly	*Carpomyia vesuviana*	Diptera	Ber	Larva	-do-
29.	Rice weevil	*Sitophilus oryzae* Linn.	Coleoptera	Stored grain	Grub and adult	Chewing and cutting type
30.	Khapra beetle	*Trogodemua granarium*	Coleoptera	Stored grain	Grub	-do-
31.	Grain and flour moth	*Siiotroga cerealella.*	Lepidoptera	-do-	Larva	-do-
32.	Pulse beetle	*Callosobruchus chinensis*	Coleoptera	Pulses	Grub and adult	-do-

Contd....

	Common Name	Scientific Name	Order	Host Plants	Damaging Stage	Mouth Parts
33.	Lesser grain borer	*Rhizopertha dominica*	Coleoptera	Stored grain	-do-	-do-
34.	Red hairy caterpiller	*Amsacta spp.*	Lepidoptera	Polyphagous	Larva	Biting and cutting type
35.	Gurdaspur borer	*Acigona steniellus*	Lepidoptera	Maize, Jawar, Sugarcane	Larva	-do-
36.	Army worm	*Mythimna separata*	Lepidoptera	Paddy, Jawar, Bajra, Wheat, Maize, Sugar.	Larva	Chewing and cutting type
37.	Shoot fly	*Atherigona varia soceata*	Diptera	Jawar, Bajra.	Maggot	-do-
38.	Fruit sucking	*Ophideres* spp.	Lepidoptera	Mango, Guava, Citrus and Pomegranate	Moth	Siphoning moth type
39.	Potato tuber moth	*Gnorinoschema operculella*	Lepidoptera Tomato, Tobacco and	Potato,	Larva	Chewing and cutting type
			Agricultural Entrance		*118*	Brinjal

Important Glossary

- *Acaricide or Miticide:* The chemical substances used to kill mites are known as acaricide.
- *Adulticide:* Chemical substance used to kill adult insects.
- *Antibiosis:* Antibiosis refers to the adverse effect of the host plant on the biology of the insects and their progeny infesting it.
- *Allelochemicals:* Allelochemicals are non-nutritional chemicals produced by an organism of one species and affect the growth, health, behaviour or population biology of individuals of another species.
- *American Bolworm:* Recently the American bollworm has become a serious pest of cotton in northern parts of India. It was first reported on cotton in 1977 in Punjab.
- *Armyworm:* The armyworm is a pest of graminaceous crops-all over the world like jawar, maize, bajra, etc.
- *Antifeedants:* The antifeedants are chemicals which inhibit or deter the feeding of insects due to their presence on the natural food of the species concerned. An antifeedant acts by suppressing the gustatory receptors.
- *Avermectins:* Avermectins are macrocyclic lactones which were originally isolated in 1976 by scientists at Merck and Co, Inc from a culture of *Streptomyces avermitilis* from Japan.
- *Allomone:* A compound released by one organism which evokes a reaction in an individual of a various species that is favourable to the emitter but not to the receiver.
- *Antimone:* A substance produced or acquired by an organism that when it contacts an individual of another species in the natural context, evokes in the receiver a behavioural or physiological reaction that is maladaptive to both the emitter and the receiver.

GENERAL ASPECTS

Each mature female just after fertilization lays about 200 to 500 eggs in a cell in which she is enclosed. The oviposition takes place into the incubating chamber which is formed by the contraction of the body of the female in forward direction inside the lac cell. The eggs are laid in the months of October and November. After six weeks of laying, the eggs are hatched into first instar nymphs in the months of November and December. When nymphs emerge they are in quite large number. This mass emergence of the nymphs is known as 'SWARMING' (Fig.).

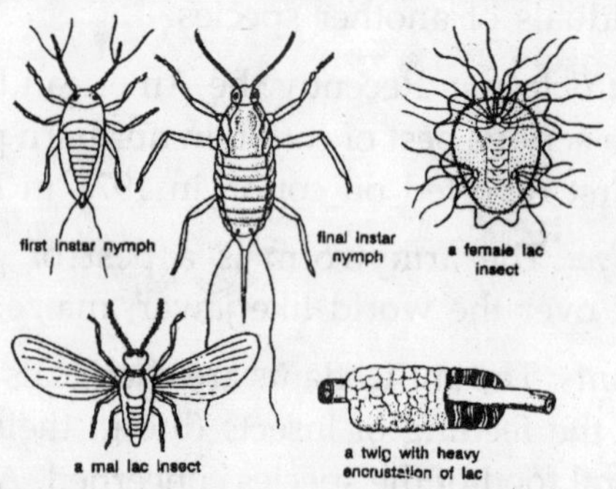

Life history of Tachardia lacca.

Nymph: At the time of emergence the nymphs are about 0.5 mm in length, red coloured and boat-shaped. The head bears paired antennae, ocelli and ventrally situated piercing and sucking type of mouth-parts. The mouth-parts are provided with proboscis. The three segmented well developed thorax contains two pairs of spiracles and only one pair of walking legs. The abdomen contains two pairs of legs and terminates into a pair of long caudal setae.

The active nymphs can crawl to a considerable distance so, just after emergence they start moving in search of food and reach their host plants, preferably on young and succulent shoots because the young nymphs are unable to settle and feed on hard twigs. These nymphs settle very close to each other on the twig of the host plant which further collapses completely and forms a continuous covering even on the lower surface of the twig. The number of nymphs that settle per square inch area is about 150 to 200. Settled nymphs suck the sap from the twig of the host plant and start to secrete the resinous substance by special dermal glands which are located all over the body. As the resinous secretion comes in contact with air, it soon becomes hard and forms a coating over the body of nymph and is called as 'CELL'. Within this cell various life processes like growth of the nymph, morphological changes and lac secretion take place.

The male 'Cell' is elongated and cigar-shaped having two holes, *i.e.*, anterior and posterior. From the posterior hole which is covered by a flap or operculum, the male insect comes out by pushing open the operculum. After six to eight weeks of stationary life the nymphs are metamorphosed as a result of which some (30%) active winged males and maximum (70%) emerge in the form of females which are wingless. The females get fixed on the host plant in resinous mass. The males walk over the encrustations of females and fertilize them within their oval cells through anal opening. The males leave the parent cell after fertilizing the female. One male is capable to fertilize many females.

The female nymph once settled never moves but undergoes 3 moults inside her cell loosing its eyes and legs, and with rudimentary antennae only. The fertilization of female is followed by a rapid growth of the female body till it begins to lay eggs in October and November. From these eggs male and female emerge in February to March. The male fertilizes the females of this generation and the fertilized female lays eggs in months of June to July and dies secreting lac all the time. Thus, the life cycle reoccurs twice in one year on the same host plant. Due to short life period males do not take major part in the secretion of lac but female secretes lac throughout her life and its life span is longer than males. Major quantity of lac is secreted from females. The life cycle period depends mainly on ecological factors of the region.

5

Silk Production

For more than 35 centuries countless generations of silkworms are continuously breeding, feeding on mulberry leaves, spinning their cocoons and dying, an everlasting sacrifice to the demand of human-beings for decoration. For the first time in 2697 B.C. Lotzu Empress of Kwang-Ti discovered the fancy origin of beautiful silk in the form of threads.

Thus, the technique of cloth preparation from the cultivated silk was known to Chinese people for more than 2,000 years ago. The art of silk preparation was kept as top secret as national policy and any one, who attempted to send the eggs of silkworm out of the country, was hanged to death. It is said that silk was as valuable as gold and ultimately gold was flowing into China from all parts of the world.

But in the year 555 A.D. this secrecy was opened and the eggs of silkworm and sericulture technique were smuggled by two monks sent as spies to China and thus, sericulture was introduced into Europe.

Later on it was introduced in Mediterranean and Asiatic countries. Nowadays sericulture has become one of the most important cottage industries in a number of countries like Japan, China, Rep. of Korea, India, Brazil, Russia, Italy and France.

The experimental and systematic study on sericulture was started in Japan in 1911 after establishment of the Sericulture Experiment Station and in 1979 the National Sericulture Experimental Station was set up in Tokyo. In India first of all Lefroy (1905-1906) started investigation on the silkworm and sericulture at Pusa Institute, New Delhi. He set up a sericulture stall in an 'All India Exhibition' organised by the Government of Uttar Pradesh at Allahabad in 1901-1910 to draw the attention of scientists and general public towards the sericulture industry.

National Sericulture Project

For the rapid development of mulberry sericulture industry in India, a five year programme. 'National Sericulture Project' of Rs. 555 crore was launched in 1989 by the World Bank and the Swiss Development Corporation under the supervision of Central Silk Board. This project was implemented by the department of sericulture in five traditional mulberry silk producing states like, Andhra Pradesh, Karnataka, Tamil Nadu, West Bengal and Jammu and Kashmir.

Twelve pilot projects in the 12 states like, Assam, Bihar, Gujarat, Kerala, Harayana, Himachal Pradesh, Madhya Pradesh, Uttar Pradesh, Punjab, Maharashtra, Rajasthan and Orissa were implemented by the Central Silk Board to create additional facilities for the silkworm seed production, training, maintenance of germplasm and extension programmes as well as research and Development facilities.

The broad objective of NSP was the introduction of mulberry sericulture in non-mulberry silk states, and expansion of this industry in mulberry silk producing states. The other important objectives of this project were generating employment opportunities for an additional one million people, increasing raw silk production, improving quality of silk produced, strengthening the infrastructure for research, extension, seed production, silk processing and development of market support for raw silk and cocoons. In addition to the above objectives the financial support to rearers, reelers and twisters, increased participation of private sectors, involvement of Non-Governmental Organisations and improved participation of women were also the priorities of NSP.

The specification of National Sericulture Project in Traditional and Pilot states was of different nature on the policy as well as application level. In traditional states the main thrust was towards consolidating the status of progress already achieved by strengthening the existing infrastructure and by providing additional facilities for further expansion. The development of water management technique, involvement of rainfed technology for mulberry, promotion of smokeless Chulhas in reeling units, establishment of grainage, basic seed farms, quality control measures, women welfare and sound Research and Development facilities were the main features of NSP.

The Central Silk Board has strengthened the already existing research facilities at various centres like Central Sericulture Research and Training Institute (CSR and TI) Berhampore, Central Sericultural Research and Training Institute (CSR and TI) Mysore, Central Silk Technological Research Institute (CSTRI) Bangalore and a number of Regional Sericulture Research Station (RSRS). Central Silk Board has also established three new research institutes like Seribiotechnology Laboratory (SBL), Silkworm Seed Technological Laboratory (SSTL) and Silkworm and Mulberry Germplasm Station (SMGS) during the tenure of NSP in traditional states.

Before the launching of the NSP in 1989 mulberry sericulture was already growing in small pockets of some non-traditional NSP states like Uttar Pradesh, Bihar, Assam, Orissa and the North- East but no integrated developmental effort were taking place in absence of infrastructural facilities. The pilot projects of CSB under NSP in pilot states provided comprehensive packages right from basic seed production to the yarn production. The CSB has invested Rs. 660 million in non traditional states from 1989 to 1994.

The pilot activities were restricted to one district (in some cases more) in each non-traditional state to ensure the achievement of the target.

The district taken under pilot project were Akola and Buldana (Maharashtra); Ambala (Harayana), Bastar (Madhya Pradesh), Dehradun (Uttranchal), Saharanpur (Uttar Pradesh); Hoshiarpur (Punjab), Jorhat and Sibsagar (Assam); Koraput (Orissa), Palakkad and Idukki (Kerala); Purnea, Kishanganj and Araria (Bihar); Solan (Himachal Pradesh); Surat and Valsad (Gujarat) and Udaipur and Banswara (Rajasthan).

Sericulture Industry and Women Welfare: It is a reality that spite of all efforts the status of women in the industrial sector is not significant because there are so many constraints for a women to participate freely in the industrial activities. Sericulture is such an agro-based cottage industry in which there is involvement of interdependent rural, semi-urban and urban based activities in which the estimated participation of women is about 60 per cent.

Thus, in contrast to other agro-based profession the role of women in sericulture industry is dominating-which will be helpful for improving the status of women in family enterprises.

In the light of women welfare through sericulture industry the Central Silk Board has established a special component of 'Assistance to Women and NGCKs' into the National Sericulture Project.

For the women welfare sericulture should lead to formulate positive additions to rural women's awareness, participation in decision making and proportional control over resources and income.

For achieving improved status and recognition of their value and importance to household, community and society as a whole, group formation should be encouraged among the women involved in sericulture industry.

Central Silk Board

Central Silk Board, a statutory organisation under the administrative control of the Ministry of Textiles, Government of India, was constituted in 1949 under an act of the Parliament. After its constitution, Central Silk Board has taken overall responsibility for the intensive development of growing sericulture industry in India. The Board is constituted by 36 members including the Chairman, Vice-Chairman, Member Secretary, representative of both the Houses of Parliament, Central and State Government's nominees, Rearers, Reelers, the Trade and Industry. The head quarter of CSB is in Bangalore.

Function of Central Silk Board

(1) Proper development of sericulture industry in the country through promoting necessary measures.

(2) To promote and encourage scientific, technological and economic research for further advancement of sericulture industry.

(3) To develop healthy silkworm seed and ensure their proper distribution.

(4) To devise advanced means for the improvement of mulberry cultivation, silkworm rearing, reeling of silk and spinning.

(5) To search and establish the means of standardisation and quality control of silk and silk products.

(6) To stabilise the prices of silk cocoon and raw silk, and rationalise the marketing.

(7) To collect the statistics regarding sericulture industry.

(8) To prepare and furnish the relevant reports on sericulture to the Central Government.

(9) To give proposal and advise to the Central Government for the intensive development of sericulture industry in the country.

Coordination of Central Silk Board with States: Being a state subject in the concurrent list, the schemes for the proper development of sericulture industry are formulated and implemented by State Governments. For performing this work, additional money is allocated to the State Governments on the basis of the annual plans approved by the planning Commission. Besides coordinating the sericulture developmental activities in states and advising the Central Government in policy making programmes, the CSB is also directly concerned with the sericulture research activities, sericulture Post Graduate Training programmes, basic seed production, distribution of raw silk, inspection of silk goods and standardisation and quality control.

The Central Silk Board has established 5 'Regional Offices' and 7 'Regional Development Offices' in different parts of the country. The Regional offices are Bangalore (Karnataka), Mumbai (Maharashtra), Calcutta (West Bengal), New Delhi (Delhi) and Srinagar (Jammu and Kashmir). The Regional Development offices are Bhubaneswar (Orissa), Guwahati (Assam), Hyderabad (Andhra Pradesh), Lucknow (Uttar Pradesh), Chennai (Tamil Nadu), Maheshmati (West Bengal) and Patna (Bihar). Special attention for promoting sericultural programmes in North-eastern region has been paid by CSB and an office of the Director (NE) is functioning at Dispur in Assam.

Silk Development and Research: Though the sericulture is one of the ancient industries in India yet it remained inadequate from technological know how and productivity.

The establishment of Central Silk Board in 1949 has been proved to be the backbone for developing sericulture industry in India which is providing the much needed research and developmental support to this growing industry. Within the span of five decades the achievement of CSB is encouraging particularly in the field of research and training programmes. To meet this requirement 'Central Sericultural Research and Training Institute was established in Mysore in 1962. Further, the status of Central Sericultural Research Station, Berhampore, established in 1943, was upgraded into a research institute.

In 1972 the Regional Tasar Research Station on Temperate Oak Tasar came into existence. The state research stations were taken over by the CSB and two Regional Research Stations were established at Titabar and Majra.

The establishment of International Centre for Training and Research in Tropical Sericulture (ICT RETS) in Mysore initiated international cooperation in sericulture development. For conducting research on silk reeling, spinning, weaving and processing of silk the first institute of its kind in the world, Central Silk Technological Research Institute,' was set up in Bangalore in 1983.

There is a wide network of extension centres carrying extensive work for the transfer of sericulture technology from lab to land through out the country. Nowadays a number of Research Institutes and Regional Research stations are actively engaged in sericulture research activities. The list of these institutes and stations is as under:

Silk-Research Institutes

(1) Central Sericultural Research and Training Institute (CSR and TI) Mysore (Karnataka).

(2) Central Sericultural Research and Training Institute (CSR and TI) Berhampore (West Bengal).

(3) Central Tasar Research and Training Institute (CTR and TI) Ranchi (Jharkhand).

(4) Central Silk Technological Research Institute (CSTRI) Bangalore (Karnataka).

Regional Silk-Research Stations: A number of Research Centres have been established in different parts of the country for carrying out research in mulberry, oak tasar, muga and eri silk:

Mulberry Silk-Research Station

(1) Bangalore (Karnataka)

(2) Chamaraj Nagar (Karnataka)

(3) Salem (Tamil Nadu)
(4) Coonoor (Tamil Nadu)
(5) Kalimpong (West Bengal)
(6) Dhule (Maharashtra)
(7) Pampore (Kashmir)
(8) Anantpur (Andhra Pradesh)
(9) Jorhat (Assam)
(10) Ranchi (Bihar)
(11) Koraput (Orissa)
(12) Mothabari (West Bengal)
(13) Majra (Uttar Pradesh)

Oak Tasar Silk Research Station

(1) Batote (Jammu and Kashmir)
(2) Bhimtal (Uttaranchal)
(3) Imphal (Manipur)

Muga Silk Research Station

(1) Boko (Assam)

Eri Silk Research Station

(1) Mendipathar (Meghalaya).

Silk Moth and its Use

The silk producing machine is an insect called as silk moth locally in Hindi 'Resham-ka-kira'. Although a number of species are found producing silk but only few species are used for sericulture industry.

Systematic Position

Phylum ....Arthropoda
Class ....Insecta
Order ....Lepidoptera
Family ....Bombycidae and Saturniidae

Silkworm Breeds

Mulberry Silkworm: *Bombyx mori.* It belongs to the family Bombycidae. China is the native place of this silkworm but now it has been introduced in all the silk producing countries like, Japan, India, Rep. Korea, Italy, France and Russia. Since the natural food of this worm is mulberry leaf, it is called as mulberry silkworm. The silk produced by this moth is white in colour.

Tasar Silkworm: *Antheraea paphia.* It belongs to the family Saturniidae and common in India, China and Sri Lanka. The caterpillar feeds on ber, oak, sal and fig plants. The cocoon produced by this moth is hard and of hen's egg size which produces relabel brown coloured silk.

Though it had been only a wild variety of silk moth since long, now, by cross breeding, it has been possible to produce such varieties which are reared any how and domesticated. But the domestication of tasar caterpillars is not so easy, so the cocoons have to be collected from the forest. The moths do not easily breed in captivity. Since breeding is not well controlled, the tasar silk industry has not reached upto mark as the mulberry silkworm industry.

Muga Silkworm: *Antheraea assama.* It also belongs to the family Saturniidae and semi-domesticated in nature. The native place of this species is Assam where, it has now become a good source of cottage industry. The caterpillars of this worm feed on *Machilus* plant and the silk produced by this moth is known as Muga silk.

Eri Silkworm: *Attacus ricinii.* It also belongs to the family Saturniidae and produces silk in East-Asia. In India sericulture scientists are trying to produce silk in East-Asia. Sericulture scientists are trying to produce such cross breeds which can provide good quality of silk and can be reared easily. It feeds on castor leaf. Cocoons cannot be reeled as in mulberry cocoons, therefore, it has to be spun. Its life history resemble with that of mulberry worms. The cocoons of this worm have very loose texture and the silk produced is called as 'Arandi' silk' locally. The threads are not glossy but much durable.

***Oak Silkworm:** Antheraea pernyi.* It belongs to the family Saturniidae and is found in China and Japan. *A. roylet* of the Himalayas and *A. yamamai* of Japan have been reared for centuries. They produce good quality of silk.

***Giant Silkworm:** Attacus altas.* It belongs to the family Saturniidae and is found in India and Malaysia. It is the largest of the living insects reaching upto eleven inches in wing-span.

Out of the above species only four are of common use for sericulture *viz., Bombyx mori, Attacus ricinii, Antheraea assama* and *Antheraea paphia.*

Mulberry Silkworm

The perusal of literature reveals that much advanced work has been done by sericulturists and scientists on *Bombyx mori* because of its total domesticated nature. Various varieties of silkworm are produced by genetical crossing and they are producing 2-7 generations in India but in some countries like Europe and Russia where the duration of winter is more than the summers only one generation is produced per year.

The race of silkworm by which only one crop is taken in one year is called UNI-VOLTINE, producing two crops in a year is called BI-VOLTINE and producing more than two crops in a year is called MULTI-VOLTINE races. Now- a-days, for commercial purpose only approved races are reared by conducting national trial of silkworm races offered by various research institutions. Different crosses are generally chosen for spring and autumn rearing.

Life History of Mulberry Silkworm: The adult of *Bombyx mori* is about 2.5 cm in length and pale creamy white in colour. Due to heavy body and feeble wings, flight is not possible by the female moth. This moth is unisexual in nature and does not feed during its very short life period of 2-3 days.

Fertilization: Fertilization is internal preceded by copulation. Just after emergence, male moth copulates with female for about 2-3 hours and if not separated they may die after few hours of copulating with female.

Egg Laying: Just after copulation, female starts egg laying which is completed in 1-24 hours. One moth lays 400 to 500 eggs depending upon the climatic conditions and the supply of food material to the caterpillar from which the female moth is obtained. The egg laying is always in form of clusters and covered with gelatinous secretion of the female moth which helps them in proper attachment.

Eggs: The eggs laid by the female moth are rounded and white in colour. The weight of the newly laid 2,000 eggs comes to about 1 gm. With the increase in time after laying, eggs become darker and darker day-by-day. Two types of eggs are generally found *viz.*, DIAPAUSE type and NON-DIAPAUSE type. The diapause type of eggs are laid by the silkworm inhabiting in temperate regions, whereas, silkworms belonging to subtropical regions like India lay non-diapause type of eggs. During diapause all the vital activities of the eggs cease.

Hatching: The eggs after ten days of incubation hatch into a larva called as caterpillar. Hatching is the most important phase of silk moth's life. After hatching, caterpillars need continuous supply of food because they are voracious feeders. If proper supply of mulberry leaf is not possible the development of caterpillar would not be in proper course. Sometimes, due to lack of food material, young caterpillars die causing great loss to the sericulture industry. It is recorded that in uni-voltine race hatching of eggs takes one month after laying.

Caterpillar: The newly hatched caterpillar is about 0.3 cm in length and is pale, yellowish-white in colour. The caterpillars are provided with well developed mandibulate type of mouth-parts adapted to feed easily on the mulberry leaves. The caterpillar is twelve segmented and the abdominal region has ten segments having five pairs of pseudo-legs. It is also provided with a small dorsal horn on the anal segment. Because of its being very much tender, the Ist instar larva can feed only on very soft leaves of mulberry plants. As they are voracious feeders, they grow rapidly which is marked by four moultings. After Ist, 2nd, 3rd and 4th moultings caterpillars get changed into 2nd, 3rd, 4th and 5th instars respectively. It takes about 21 to 25 days after hatching.

The full grown caterpillar is 7.5 cm in length. It develops salivary glands, stops feeding and undergoes pupation. The time taken for the full growth of the caterpillar from young to the well grown stage varies with regard to the temperature, humidity, food supply and type of race. Weight of the full grown caterpillar varies from 4 to 6 gm.

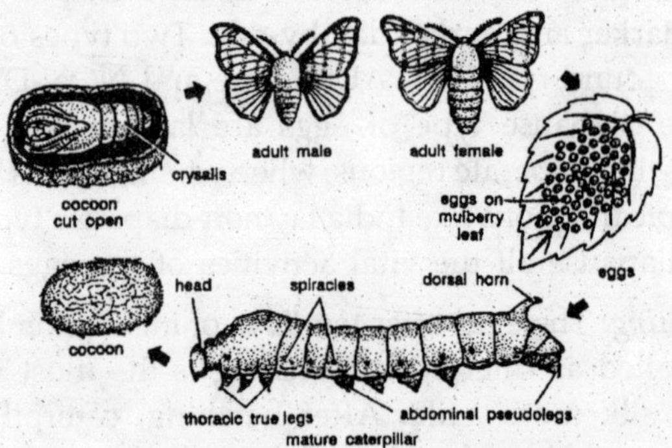

Life history of Bombyx mori.

Pupa: The caterpillars stop feeding and move towards corner among the leaves and secrete a sticky fluid through silk gland. The secreted fluid comes out through spinneret (a narrow pore situated on the hypopharynx) and takes the form of long fine thread of silk which hardens on exposure to the air and wrapped around the body of the caterpillar in the form of a covering called as COCOON.

Cocoon: Cocoon is the white coloured bed of the pupa whose outer threads are irregular while the inner threads are regular. The length of continuous thread secreted by a caterpillar for the formation of cocoon is about 1000-1200 metres which requires 3 days to complete. The threads is wound around the cocoon in concentric manner.

The binding of threads round the cocoon is very interesting and quick going phenomenon achieved by the constant round motion of the head of the caterpillar from one side to the other at the rate of 65 times per minute.

Now the silkworm pupa is covered within a thick, oval white or yellow silken cocoon. It is estimated from the data obtained by practical application that one pound of silk can be obtained from 2,500 cocoons. The weight of one cocoon is about 1.8 to 2.2 gm and the weight of the cocoon shell only is 0.45 gm. The size of the thread is 2.0 to 2.8 denier. The pupal period lasts for 10 to 12 days and the pupae cut the cocoon and emerge into adult moths.

Emergence of Imago: Due to active metamorphic changes during pupal period the abdominal pseudo-legs disappear and two pairs of wings develop. The silkworm within the cocoon secretes an alkaline fluid to moisten its one of the ends. As a result of this the moistened end becomes soft where the threads are cut open (hydrolysed) by the silkworm. Finally a hole is formed through which a feeble adult moth squeezes out of the cocoon.

SERICULTURE

The production of silk from the silkworm by rearing practices on commercial scale is called sericulture. Although silk is very beautiful and fancy gift of nature but its commercial production is very much complicated and a tough job which requires heavy man power. Before going into the details of the sericulture industry it is essential to have an idea about the major, steps and requirements of sericulture. Improved race of *Bombyx mori* and good nutritive type of mulberry plantation is the sole need for the industry. Other than these some of the other requirements are given below:

(1) *Machana:* The proper place for rearing the silkworms.

(2) *Rearing trays:* For keeping the laid eggs alongwith the mulberry leaves.

(3) *Spinning or Chandrakis tray:* For keeping the caterpillars at the time of full grown stage ready for pupation.

(4) *Dalas:* For fetching the mulberry leaves.

(5) *Chopping knife:* For cutting the mulberry leaves into smaller pieces.

(6) *Baskets:* For the distribution of mulberry leaves.

(7) *Hygrometer:* To know the % humidity in the atmosphere.

(8) *Thermometer:* To take the reading for the room temperature.

(9) *Oven:* To regulate the different stages of life cycle at different temperatures.

(10) *Freez:* For the storage of eggs (seed) for next generation.

Selection of Races of B. Mori: Although a number of races have been developed for the purpose, care should be taken to select only those which would be fit for a particular locality keeping in view the climatic condition of that area.

According to rearing season *viz.*, spring, early autumn and late autumn different crosses are chosen for successful rearing.

Mulberry: The whole sericulture industry is based on the best utilisation of the mulberry leaves as it is the only food of this insect. In Japan more than thousand varieties of mulberry exist including as many as 130 natural species out of which only a few are in commercial use.

Mulberry is a deciduous medium sized tree belonging to the family Moraceae. Its species are given as below:

Morus Alba Linn: It is of white type and locally called as Toot, Tootri etc. This is widely used for rearing the silkworm and also for fruits.

Morus Indica Linn: It is locally called as 'shahtoot' and is Indian in origin. It is also used for fruits and for rearing of silkworms.

Climate: Although, mulberry is a plant of temperate zone but some varieties have been developed by cross breeding which can grow well in tropical and subtropical zones also. According to the climatic conditions different varieties of mulberry are being used throughout the world.

Soil: For a good crop and nutritive type of leaves, soil should be of loam type and not with stones etc. Soil with neutral pH is best suited for proper growth of mulberry plants.

Plantation: It is of two types as given below:

Seed: In December and May seeds of mulberry are sown and the seedlings grow very quickly which can be transferred either as whole plant or a stump in months of June, July and November. Seed beds should be mixed with ash, lime and white arsenic.

Cutting: Cutting should be done in months of July and August. The size of each cut piece should be 20 to 25 cm in length and 2 to 3 cm in thickness. The cut pieces should be planted in slanting position in the pits (50 × 50 × 50 cm) having 1 kg. superphosphate, 30 gm B.H.C. and 25 kg farm-yard manure.

The plantation is performed in December-January or July. The plantation is mainly of 2 types *viz.*, single row and multiple row plantation. (i) In single row plantation 400 plants can be placed around the boundary of one hectare plot at the distance of one metre from plant to plant. (ii) In multiple row plantation scheme the whole of the plot is used only for sericulture programme. About 10,000 plants are placed at the distance of 1 × 1 metre per hectare. Weekly irrigation is most preferred for bumper growth of mulberry leaves.

Fertilizers: The months of December to January are the best time for manuring the crops. Although the real dose of any fertilizer can be fixed after proper soil testing and according to the fertility of the soil but a generalised recommended dose of manuring is given below (S.S. Sagwal and O.P. Katna, 1981).

Dose of Manures and Fertilizers per Plant

Age of Plant (Year)	*Farmyard manure (Kg.)*	*Kisan Khad (gm)*	*Super phosphate (gm)*
1	10	500	250
2	20	1,000	500
3	30	1,500	750
4	40	2,000	1,000
5	50	2,500	1,250

Pruning and Thinning: December and January are best suited months for pruning and thinning. After 4 to 5 years of plantation mulberry plants get ready for the rearing of silkworms.

The white mulberry *(Morus alba)* or Kashmir white or black mulberry or white Philippine early variety are preferred. In India white Philippine early variety is most preferred because this grows easily.

Production of Silk

The word rearing does not mean only the feeding of caterpillars as often understood but a continuous care from egg laying through aestivation, hibernation, incubation, early stage larval care, late stage larval care to the production of cocoon. So, for proper and step-wise study of rearing one should proceed from Grainage Technology.

Grain Management

The aim of the establishment of grainage is to provide good quality of seed to rearers and maintenance of original quality of races. For this purpose due care should be taken of the 'crop of silkworm' for seed production from the very beginning *i.e.* the caterpillar stage, by providing them with proper nutrition and protection from the attack of diseases.

Keeping these points in view initial selection is made on the basis of percentage of dead pupae during normal development. If it is above the limit seed should not be purchased for seed production. First selection is made by separating out dead cocoons and next selection in the grainage.

After final selection, cocoons are subjected to sex separation by cutting one end of the cocoon either manually or through Nagahara (in Japan) cocoon openers. The Nagahara machine can cut 10,000 to 15,000 cocoons per hour. For the production of commercial eggs loose forms of cocoons are used and collective mother examination is done, either through mass pebrine detecting machine or general microscopic observations. They are then kept for mass emergence.

Emergence of Moth and Fertilization: When kept for emergence at room temperature, mass emergence of adults takes place. As per their nature, just after emergence, male moth starts moving around the female.

Males are very much active whereas the females which are loaded with eggs are incapable of flying. If not separated at once in cages males start copulating with the females but the eggs obtained from this female mated from the male of the same stock is useless for the seed.

So the males and the females just after emergence have to be separated into separate cages without their mating. Now one female of one lot is kept with the male of the other lot and atones they form pair and copulate for about 3 hours. After completion of mating, males should be separated and may be used for the fertilization of other females. But one male can not fertilize more than two females. Now fertilized females are subjected to egg laying.

Egg Laying: Just after fertilization, female starts egg laying and in the duration of 24 hours it completes egg laying process. The eggs laid by one female are about 400 to 500 varying according the different races. Female dies after egg laying. These eggs are called as SEED. These eggs are kept in sterilised trays and stored at 4°C under laboratory conditions or sometimes kept at hill stations in diapause conditions.

The stages of egg (seed) production is of 3 types *viz.*, production and supply of grand parent eggs, production and supply of parent eggs, rearing of parent eggs and production of commercial F_1 seed. The grand parent and parent eggs are produced by recognised and reputed organisations. This commercial seed is supplied to the rearers.

Hatching: This is an important phase of sericulture industry because as soon as the larvae are hatched they start feeding voraciously. So only those sericulturists who would be able to supply sufficient amount of fresh mulberry leaves to the young hatched larvae, could perform successful sericulture programme otherwise young ones will die resulting great loss to sericulture industry. This is why the hatching has to be controlled, accelerated or postponed by artificial treatments under refrigerated conditions.

For proper hatching of seeds (eggs) advanced techniques have been developed in which eggs are collected and kept with mulberry leaves, working as stimulant for hatching in shady places on white sheet of paper in insect proof trays on a stool. For this purpose the legs of stool must be kept in water so that insects may not crawl and damage the hatching eggs.

It is also notable that if the eggs are placed in the same position in which they are laid, hatching will not be 100 per cent. So it is advisable that tne eggs kept in trays should be moved with the help of feather. The group of caterpillars hatched at various stages should be kept separately. Thus, one should be careful that hatching must be coincided with the best season of the mulberry.

Experimental Data of Egg Laying and Hatching: (i) If after 120 minutes of oviposition eggs are kept at 10° C for 24 hours then transferred to the oven at 16°C for 4 days and further soaked in hydrochloric acid (15% at 46° C) for 5 minutes, the diapause condition is broken easily. (ii) To get the homozygous stock of female silk moth, eggs are taken from 24 hour old moth and treated with hot water (46° C) for 18 minutes then kept at 15 to 17° C for 4 days, treated with acid at 46° C and then kept for incubation. When such eggs are reared carefully a homozygous stock is obtained. (iii) If the female moth is kept at 5°C for 24 hours just after mating and for egg laying at normal room temperature, 70 per cent eggs are laid in the first hour after incubation and 20 per cent eggs are laid in 2nd hour. This data is very much useful for grainage technicians from the point of view of uniform hatching of eggs.

Rearing as a Part of Trading

After grainage management the next step is the supply of seed or caterpillars to the fanners. The supply is of two types depending on the knowledge of rearers *i.e.*, supply of eggs and 2nd instar larvae. The old rearers who are well versed with the rearing technique may purchase eggs for the rearing but new and untrained rearers knowing nothing about the rearing should always be given 2nd instar caterpillars for this purpose. Much care should be taken for the rearing of 1st, 2nd and 3rd instar caterpillars and 4th and 5th instar caterpillars are mostly reared either on hanging trays often with nylon nets or on the floor.

It is essential to clean the bed of the caterpillars once in a day for 3rd, 4th and 5th instars. The principle of rearing should be the production of healthy caterpillars with uniform development. The mulberry leaves supplied to the 1st and 2nd instar larvae should be young and well chopped because they are very tender in nature. The 3rd instar larvae may feed even on chopped hard leaves but 4th and 5th in star larvae can feed easily on leaves also.

The taste of mulberry leaf is lost due to rapid loss of moisture, so rearer should be careful to stop or reduce -the loss of moisture from the leaf after harvesting. For this purpose 60 to 70% humidity of the room is needed otherwise it affects the health of silkworm.

So, it is advisable to keep the leaves wrapped in wet clothes. The overcrowding of worms causes under-nourishment and checks proper development. If a large number of rearers in the same village take the worms from same lot, spinning of silkworm of all the rearers starts on the same day if proper supply of mulberry leaves is there.

Now full grown 5th instar larva stops feeding, undergoes pupation and settles to. a corner among the mulberry leaves and starts to secrete sticky fluid. Thus, an improved technique of rearing of silkworm has resulted in the production of cocoons of good quality. The temperatures best suited for the rearing of 1st, 2nd, 3rd, 4th and 5th instar larvae are 27, 27, 25, 24 and 23° C respectively.

Spinning of Cocoons

This is the period when the caterpillar stops feeding and starts to secrete a pasty substance from the silk gland. In this condition worms should be picked up and transferred to the spinning trays and kept in a position of slope (slanting) to the sun for a short period. Within three days spinning is over and the cocoon is formed and this is the last phase of the rearing of silkworm.

Quality of Cocoon: The quality of cocoon is dependent on the raw silk yield, filament length, reliability and splitting.

Marketing of Cocoon: The price of cocoon is fixed during every season of the rearing. This price is, however, watched by the Government and cocoons are purchased by the rearers.

Threading

The method of obtaining silk thread from cocoon is known as post-cocoon processing. This includes STIFLING and REELING.

Stifling: The process of killing the cocoons is termed as stifling. Sericulturists should be very much careful that before the emergence of silkworm (The cocoons which do not have cut holes) good sized cocoons of 8 to 10 days old are selected for further processing and dropped into hot water or subjected to steam or dry heat, sun exposure for 3 days or fumigation. In this way pupae or cocoons are killed.

The killing of the cocoon in boiling water helps in softening the adhesion of the silk threads among themselves and loosening of the outer threads to separate freely, facilitating the unbinding of silk threads.

Reeling and Spinning: The process of removing the threads from the killed cocoon is called as reeling. Four or five free ends of the threads of these cocoons are passed through eyelets and guides to twist into one thread and wound round a large wheel from which it is transferred to spools. Thus the silk obtained on the spool is called as RAW SILK or REELED SILK. The waste outer layer or damaged cocoons and threads are separated, teased and then the filaments are spun. This spunned silk is called as 'SPUN SILK'.

The raw silk is further boiled, stretched and purified by acid or by fermentation and then carefully washed over again and again to bring about the well-known lustre on the thread.

The modernisation of the reeling and spinning process by autolisation and various labour saving process has opened a new way to this cottage industry in the world. One autoliser can yield 3.715 kg silk per basin in 8 hours in Japan.

Silk: Silk is a pasty secretion of the silkworm produced by the silk gland. The silk glands are actually modified salivary glands which are long and sac like. As this pasty secretion comes in contact with air, it becomes hard and forms strong and pliable silk strands.

This secretion forms two cores of fibroin: (i) a tough elastic insoluble protein consisting of 75% of the fiber's weight and cemented together with sericin from the middle region of the silk gland at the time of secretion, and (ii) a gelatinous protein which is easily soluble in warm water.

Some quantity of wax and carotenoid pigments are also detected. The diameter of the silk fibers is 0.0045 to 0.0082 cm. Its elasticity is found to be 20%.

Diseases of Silkworm

The Sericulture Industry suffers from a number of diseases in Tropical regions of South East Asia. The maggot disease, pebrine, polyhedrosis and flacherie are the diseases which cause severe damage to this industry. The poisoning by tobacco and Muscardines is also reported to be harmful but is not very common.

Harm of Maggot Disease

This disease is caused by *Tricholyga sorbillans*, a fly belonging to the order-Diptera and Family-Tachlnidae. It is distributed throughout India, Japan, Korea, China, Vietnam and Thailand. The presence of milky white cylindrical eggs on the skin of silkworm larvae is a symptom of this disease. The number of eggs laid by the fly on a silkworm larva varies from several to more than fifty but usually they are two to three. After 30-40 hours of egg laying, maggot is formed inside the egg shell. Now this maggot makes a hole on the ventral side of egg shell and on the skin of the silkworm.

Thus, maggot penetrates into the body of silkworm larvae and starts eating the tissues of larvae. When the maggot penetrates into the larval body, the big black mole is formed on that part of the skin. The segments in which maggot exists, swell up and bend as a result the attacked larva becomes inactive and loses appetite. Usually fourth and fifth stadia are attacked by the maggots. The larvae, attacked upto the fourth stadia die before making cocoon, whereas, those attacked in the fifth stadium make cocoons but usually do not attain the pupal stage.

A number of natural enemies like parasitic insects, fungi etc. are known to control the population growth of *T. sorbillans*.

The avoidance of entry of this fly into the rearing room of silkworm is the only means for the prevention of this disease for which nets are fixed around the windows.

Definition of Pebrine

This is one of the worst diseases of silkworm. Sericulture was once damaged by this disease in all the countries involved in this industry but some countries have overcome this disease and succeeded in getting pebrine-free-silkworm-eggs for reeling cocoons. *Nosema bombycis* Nageli, belonging to Microsporida, is the casual micro-organism of this disease.

The infection takes place through the mouth of larvae at the time of feeding or through the mother's ovary. When new spores are formed in the tissues of alimentary canal of silkworm, they are discharged with the faeces and make a source of infection.

When the spore enters into the digestive tract of silkworm, two nuclei, contained in the sporoplasm, are divided into four nuclei and at the same time the polar filament projects and penetrates into the cells of the alimentary canal.

Further, they enter into the blood and swim in it. They are distributed throughout the host body, attacking various tissues, specially the fat bodies and organs, excluding chitinous tissues and nucleus of cells. When the hypoderm is attacked and its cells die, the affected part becomes black due to the formation of melanin.

Inside the body, the milky white spots or marks are observed on the silk gland or on the surface of the alimentary canal. In case of severe infection of the eggs the whole of yolk nearly gets filled with the micro-organisms resulting in their (eggs) death. Whereas, in case of slight infection the eggs hatch but the larvae carrying infection die at the third moult without making cocoon.

The diseased larvae show little inclination towards food

and exhibit irregular and differential growth resulting in the formation of small larvae of different sizes. They become tardy, shrunken or moult quite late and finally die.

Counter measures against pebrine. If the infection occurs at the embryo stage, the larvae die in the third moulting stage as already mentioned. So, during the rearing of reeling cocoons it is very important that the pebrine free eggs are used.

The infection by pebrine should be strictly avoided at any stage of the life history of the silkworms for rearing seed cocoons in order to get the healthy eggs. Therefore, it is essential for the prevention of pebrine that the tools used and the rearing houses should be kept free from pebrine germs. The following measures may be applied.

Examination of Mother Moths: All the mother moths producing eggs should be carefully examined one by one, and eggs laid by healthy moths only be taken for further use. The unqualified eggs should be burnt away.

Forecasting and Correcting Examination: In order to make the pebrine examination more reliable, the forecasting and correcting examinations are carried out for the eggs of the seed cocoons. The materials employed for forecasting examinations are the excrement of mature larvae, late moulting larvae, dead larvae, cocoons or pupae and moths accelerated to emerge out of cocoons. For the correcting examination a few eggs of each reproductive egg-batch are taken and incubated and thus, emerged larvae are used as the material for examination of pebrine germs.

Removal and Disinfection of Pebrine Germs: The spores of pebrine may survive for a number of years in an ordinary type of rearing room if the environmental condition is humid. Therefore, rearing rooms, tools and other utensils should previously be cleaned and washed to remove the pebrine diseased eggs, carcases of infected larvae, pupae, moths and dead cocoons, faeces of diseased larvae and so on. The pebrine spores can be destroyed by treatment with 2% formalin for

30 minutes, 0.5% sublimate for 5 minutes, 5% chlorinated lime for 30 minutes, current steam for 30 minutes and sun shine in the summer for 7 hours.

Care on Rearing Silkworms: The tendency of the infestation by the pebrine spores has been observed to occur more when silkworms are reared in dry and cool conditions than in the hot and wet conditions. If the larval period becomes longer, the infection of pebrine is severe, so care should be taken on these points during the course of selection of the seed cocoons by the rearers.

Silkworms and Polyhedrosis

Three types of polyhedrosis are found in silkworms:

(1) Nuclear polyhedrosis.

(2) Intestinal cytoplasmic polyhedrosis.

(3) Intestinal nuclear polyhedrosis.

1. *Nuclear Polyhedrosis (Grasserie, Jaundice):* It is caused by a kind of virus which forms polyhedra in the nuclei of the cells of fatty tissues, dermal tissues, muscles, tracheal membranes, basement membrane, epithelial cells of midgut and blood corpuscles. The polyhedra are commonly hexagonal and rarely tetragonal in shape, containing large number of virus in them. The polyhedra form the white pus after they are released into the body fluid. The virus present inside the polyhedra maintains its pathogenic power for a number of years in the rearing room but the isolated virus loses its pathogenic power in a period of short time.

The larvae infected by this virus become inactive and lose appetite, and the membrane between segments swell up. Further, the whole body swell up showing the loose skin. In the last phase of the disease the body becomes purulent and the skin becomes tender from which pus leaks out. The larvae in this phase crawl around up and down and finally die. The infection of grasserie occurs through the mouth of the larvae, the wounds of skin and induction under extremely adverse conditions (cold treatment, heat treatment, chemical treatment).

Countermeasures

(1) The rearing rooms, tools and utensils should be washed, cleaned and disinfected.

(2) The highly nutritive mulberry leaves should be given sufficiently to the infant larvae. The fresh air should be circulated in the rearing room especially in fifth stadium.

(3) The high temperature, low temperature and high moisture content should be avoided especially during the infant stages to maintain good health.

(4) The diseased larvae or dead bodies should be removed and the room disinfected completely by 2% formalin.

(5) The silkworms should be reared carefully, so that they may not suffer from the wound of skin.

Intestinal Cytoplasmic Polyhedrosis: This type of polyhedra are formed in the cytoplasm of midgut cells but in a few cases they are formed in the goblet, too. The polyhedra of this disease contains plenty of virus. The infected cells of midgut rupture and polyhedra are released into the gut. Thus the faeces, excreted becomes whitish, containing plenty of polyhedra.

The symptoms of this disease are the occurrence of translucent cephalothorax and shrinkage of body size but in the advanced phase the larvae excrete white coloured loose faeces. The body fluid of the diseased larvae remains in the normal condition.

The mode of infection of this disease and countermeasures are similar to those of the nuclear polyhedrosis.

Intestinal Nuclear Polyhedrosis: In this disease the virus makes polyhedra inside the cytoplasm and nucleus of the midgut cylindrical cells. The polyhedra formed in this disease are large sized. The translucent cephalothorax, shrinkage of body and diarrhoea are the symptoms of this disease also. The mode of infection and countermeasures are similar to those of nuclear polyhedrosis.

Role of Flacherie

Flacherie is the genetic name of some kinds of silkworm diseases, carcases of which rot due to the attack of bacteria. Flacherie may be divided as follows:

Infectious Flacherie caused by a Kind of Virus: The various symptoms of flacherie are loss of appetite, translucent cephalothorax, vomiting and diarrhoea but the real diagnosis can be made after the microscopic observation of the virus.

The pathogen of this disease is a spherical virus which does not form polyhedra in the body of silkworm larvae. The virus multiplies in the tissues of midgut and is released into the gastric juice and is excreted alongwith faeces which is the source of infection. The virus infects the larvae of silkworm orally.

Countermeasures

(1) Resistant silkworm races should be selected.
(2) The rearing room, tools and utensils should be well disinfected.
(3) In order to maintain the good health of silkworm, high quality mulberry leaves should be provided.
(4) Favourable conditions like temperature and humidity should be maintained in rearing rooms.
(5) The faeces, diseased larvae and dead bodies should be piled in the compost.
(6) Some of the chemicals like hydrochloric acid, formalin, chlorinated lime may be used as disinfectants for this virus.

Gastric injury caused by physiological disturbance of silkworms followed by the multiplication of bacteria. Due to the supply of bad quality of mulberry leaves the digestive physiology of the silkworm is disturbed and multiplication of bacteria in the gastric cavity takes place.

Thus, the combined action of physiological disturbances and bacterial activity in the gut are major causes of this disease. In unfavourable climatic conditions, the bacteria like *Streptococci* sp. *Coli aerogenous bacilli* or proteus group bacilli attack the weakened silkworms. The control measures against this disease is to keep healthy conditions of rearing silkworms.

Bacterial Intoxication: This disease is caused by a toxin of some bacilli, *Bacillus thuringiensis* Var. The larvae attacked by this toxin become unconscious, later soften, become darkish and finally rot off.

The infection occurs orally and can retain the toxicity as long as for seven years in some cases. The countermeasures are the disinfection of the rearing room and instruments.

Septicaemia: This disease is caused by infection of some bacteria as *Bacillus megatherium, B. proteus, B. prodigious, B. pyocyones* in the blood of silkworms. This disease is rare in the larval stage but it causes severe damage to the pupae and the moths during the period of egg production. The infection is caused through the wounds on the skin.

Countermeasures

(1) The rearing tools and rearing houses should be kept sterilised.
(2) The diseased cases should be kept away and put into the fermenting compost.
(3) During the period of egg laying the temperature and humidity of the room should be maintained in proper order specially at lower range of humidity.

Green Muscardine in Use

It is a fungal disease of silkworms. There are a number of muscardine in silkworm but only green muscardine *(Spicaria prasina)* has been noticed to affect the larvae of silkworm in

Vietnam. The infection may be observed at the third and fourth stadia of silkworm. In the beginning stage a big black spot is observed on the ventral side.

The green-tubes of fungus develop into mycelia in the blood and bear cylindrical spores which are separated from mycelia and further form mycelia which bear cylindrical spores again. Thus, all the organs of silkworm are attacked by this fungus disturbing their normal functioning. As a result the diseased larvae do not moult and finally die.

Countermeasures

(1) The rearing room, tools and utensils should be disinfected.

(2) The diseased larvae and their faeces should be piled into compost to kill the germs.

(3) The rearing bed should be kept dry as much as possible to prevent the germination of conidia of fungus.

Economy as an Operating Resource

The raw silk is used in the manufacture of woven materials and the knitted fabrics for the preparation of garments, parachutes, parachute cords, fishing lines, sieve for flour mills, insulation coil for telephones and wireless receivers, and tyres of racing cars. Fabrics for garments in various weaves, plain, twill, stain, crepe, georgette and velvet, knitted goods such as vests, gloves, socks, stockings, dyed and printed ornamented fabrics for saries, jackets, shawls and wrappers are made out of this material.

Sericulture as Trading

At the root of the social, economic, cultural and political progress of India, there are 6.5 lacs villages where 75% of the population of the country lives. The real progress of our country is definitely on the development of these villages. For the economic independence of the villages it is most important to check the flow of ingenious people from the villages towards cities so that the villages also may get a chance for advancement and progress. Even today the main source of earning livelihood, in villages, is agriculture and agriculture based industry on which 70% of the population depends.

Majority of the village population even today lives below poverty line. It is believed that small scale and cottage industries are only in small and under developed countries but this assumption is only partially correct because in developed countries like USA, Germany, France, Britain, Russia, Switzerland and Denmark cottage industries occupy a special importance in their economy. In such countries where there is derth of capital but ample labour power, small scale and cottage industries also have great importance. When we discuss the economic policy of India, one doubt always haunts our mind, weather on the strength of small and cottage industries only, would it be possible for us to face the challenges in the international economic competition. Gandhi's economic philosophy provides solution to this doubt that small cottage industries and modern and large industrial units have to be free from mutual competition, then only, all round development of the nation can be possible.

In the back ground of the industrial growth of past few years and their popularity and distribution of cottage industries it is apparent that the silk industry has developed as a popular cottage industry in which all big, middle, marginal and small farmers in their own way in their limited resources can start silkworm rearing. Even the landless labours with the cooperation of farmers can get mulberry leaves and rear silkworms. This industry neither needs any large equipment, big capital nor specialised technical know how for its start. Although sericulture is practised in India since long even much before independence, yet because of more capital involved at that time and low quality production it could not achieve required success.

In any region for achieving success in sericulture industry there are some basic factors which should be surveyed. It is most important to have additional knowledge of the climate, soil, previously developed small industry, if any, and economic and social structure of that region. In case the soil and the climate of the region is suitable for rearing of silkworms, proper arrangement for irrigation, availability of labour etc. have also to be thought of. In short it must be fully ascertained weather this new industry of silkworm rearing would be definitely profitable in contrast to the traditional farming or other small and cottage industries, if any, prevalent in that region before embarking on this new project.

The idea of the capital investment involved and the average income per acre should also be calculated so that the farmer be told that it would be profitable to them. If finally decided to undertake this venture then for selecting the race of silkworm for rearing, the climate, soil and irrigation facilities should again be kept in mind so that the rearers may not face any technical difficulty later on.

In the country due to the efforts of sericulture research and investigation cum training centres and other production programmes, there has been a good improvement in the production and quality both of the silk cloth during the last 25 years. In a number of regions new varieties of mulberry with close spacing and proper instructions have proved to be much fruitful in some regions but a number of rearers are still unaware of the proper cultivation technique and use of fertilizers for the mulberry plants. By efforts of systematic researches it has been possible to improve the nutritive quality and growth of mulberry leaves from 15,000 to 35,000 kg per hectare.

The new mulberry varieties K_2 and M_5 are giving 65% more yield than the local crop. Now some strains are capable of giving 100 per cent increase in the yield of mulberry leaves. Sericulture research stations and grainages have evolved techniques for handling of young leaves easily with respect to temperature, humidity, quality, quantity and frequency of feeding, spacing, aeration and protection against diseases. A number of techniques have developed which has resulted in the improvement of the maintenance of silkworms.

Although India occupies second position after China in silk production yet the production in India is one fourth of the total production in China. The increasing demand of silk cloth in India and its decreasing production in Japan has placed China as the biggest market of silk in the world. The ever growing interest shown by Japan in the manufacture of electronic goods has left Japan much behind in silk production. Central Silk Board was established in 1949 in India and since then, through development and research plans, silk industry got a great encouragement.

But in 1989, with the cooperation of 'World Bank', 'National Silk Project' has doubled in silk production in seven years and in the next five years the production may be still doubled from the present quantity. 85% of the silk cloth manufactured in India is utilised in the country itself. India imports raw silk for its industry from China. Out of the total silk cloth woven here, 65% is woven by handloom, 30% by powerloom and only 5% in mills by modern machine.

In India the export quality of silk can only be produced by using threads manufactured by modern machines only. At present 3.5 lacs hectare of land is under mulberry cultivation in India. In Karnataka, Andhra Pradesh and Tamil Nadu the interest of the farmers in silk production is continuously increasing day by day. In the country to day silkworm rearing programme is providing whole time and part time employment to 75 lacs of people.

Lift-up of Silk Industry

Of the 6.5 lac villages in India sericulture is practised in about 75,000 villages. The production of mulberry silk is confined to the five traditional states like Andhra Pradesh, Karnataka, Tamil Nadu, West Bengal and Jammu and Kashmir. Now it has also been initiated in other twelve states like,

Kerala, Maharashtra, Bihar, Madhya Pradesh, Orissa, Gujarat, Rajasthan, Punjab, Harayana, Uttar Pradesh, Assam and Himachal Pradesh and gaining success in some other states.

The traditional states produce about 99 per cent of the total mulberry raw silk in the country. The major production of tasar raw silk is achieved from Bihar, Orissa, Madhya Pradesh, Andhra Pradesh and West Bengal, while Maharashtra and Uttar Pradesh are producing tasar raw silk on a small scale. The eri raw silk production is being achieved from major eri silk producing states like Assam, Bihar, Meghalaya and Manipur while it is also produced in Arunachal, Mizoram, Tripura, Nagaland and Orissa at small scale.

Manipur, Nagaland, Mizoram and Arunachal Pradesh are the oak tasar silk producing states. The production of muga silk is mostly confined to the states of Assam but nowadays this silk is also produced in Mizoram, Meghalaya and Nagaland.

6

Honey Production

Man had started the use of animal products since time immemorial even at the cost of animal's life. Honey has been under use in human civilization since prehistoric period as mentioned in our religious literatures like Vedas, Puranas, Ramayana, Mahabharata and Charak Sanghita. Some foreign travellers like Fahiyan and Huen-T-sang had also discussed the use of honey as medicine. People were very much dependent upon honey for medicines and essential nutritive elements of the diet.

The highly evolved social organisation of bees had been established before the existence of human race. Bees teach us the lesson of work-work and work with cooperation. We can easily imagine about the hard-working of the bees by the simple fact that for one pound of honey, a single honey bee travels about double the distance of the circumference of the earth's globe.

Previously, the method of extraction of honey from the honey comb was very much crude but after the invention of artificial hive by Longstroth (1951), it became scientific and commercial.

The bee keeping in United States, Canada, Australia and New Zealand has achieved out-standing success. In India people do not take interest in bee keeping from commercial point of view but only for their routine use. Since honey is produced by the honey bees, a detailed study of the biology of the bees is essential for successful implementation of apiculture programme.

Honey Bee

Phylum .... Arthropoda

Class .... Insecta

Order .... Hymenoptera

Family .... Apidae

Genus .... *Apis*

Flora and Apiculture

Although honey bees can collect nectar and pollen from quite a long distance but the flora for apiculture is also important. The flora may be of wild or cultivated type. The more nectar yielding plants are neem, jamun, soapnut etc. The other plants like maize, rose and sorghum are good sources of pollen. Some plants like plum, cherry, apple, sheesham, coconut, guava, mustard etc., are good sources for nectar and pollen both.

Selection of Bees

For running a good apiary, selection of honey bee is of much importance, so the following should be kept in mind at the time of selecting honey bees for apiculture:

(1) Honey bees should be of gentle temperament.

(2) Honey bees should have capability to construct strong colony.

(3) It should have ability to protect from enemies.

(4) Honey bee should have energetic and industrious workers.

(5) Workers can suck juice from numerous varieties of plants.

(6) Bees on the whole can produce more and more honey from its comb.

(7) Bees can from their comb easily at any place.

The apiculture scientists engaged in genetics are trying to find out such cross races which would not be of ferocious nature but be a good honey producer. In India, *Apis indica* is the best bee for apiculture industries due to its gentle nature and having efficient and prolific workers.

Honey Bees: Classification

Four species of honey bees are reported:

Apis Dorsata F. (Rock bee): This the largest bee, about 20 mm. in length, so named as GIANT HONEY BEE. SARANG and BOMBARA are other names of this bee which yields maximum amount of honey in comparison to other species. A single comb may yield 60 pounds of honey which is the maximum amount for a comb. The workers are very smart and active which may pollinate 12,000 flowers daily. But due to its ferocious and irritable nature, specific hive and migratory habit it is very difficult rather practically impossible to domesticate them for the bee keeping industry.

Apis Indica F. (Indian bee): Commonly found in forest and plain regions of India. This is slightly smaller than *A. dorsata.* They prefer to live in dark places and construct several parallel combs about one foot across the protected places like cavities of tree trunks, mud walls, earthen pots, thick bushes, wells and walls of the buildings. This species is very gentle in nature, so can be domesticated easily. The production of honey is much less *i.e.,* 6 to 7 pound per comb.

Apis Florea F. (Little bee): This is smaller than *A. indica* and yields very small amount of honey. The bees are not of gregarious nature and form a single comb. Due to its docile nature and rare stinging behaviour the combs can be removed easily for the honey extraction.

Apis Mellifera F. (European bee): Although this bee produces less honey yet it is found to be the best species from the commercial point of view. Due to their docile nature they can be domesticated easily and can be improved by breeding for several hundred years. Out of its several varieties, the Italian variety is reared everywhere in Europe and America in artificial hives for honey.

Establishment of Society

A highly organised division of labour is found in the colony of honey bees. A good and well developed colony of bees had 40 to 50 thousand individuals consisting of 3 castes *viz.*, QUEEN, DRONE and WORKER.

The queen after fertilization lays fertilized and unfertilized eggs both. From unfertilized eggs male bees emerge which are known as DRONES whereas from the fertilized eggs worker bees are produced. The workers when feed on ROYAL JELLY, develop into QUEEN.

Queen: It is a well developed fertile female provided with immensely developed ovaries. Commonly one queen is found to be present in each hive and feeds on Royal Jelly. She is the queen in real sense as the Mother of the Colony, guarded by a number of attendants and never allotted any duty except egg laying. Egg laying is the sole function of the queen throughout her active life span.

The queen is 15 to 20 mm in length and can be easily distinguished by her long tapering abdomen, short legs and wings. Structurally she is unable to produce wax or honey or gather pollen nector. By the combination of ovipositor-cum sting, a structure is developed which aids in egg laying. It is said that the queen gets mated only once in her life but in a single chance of mating, drone releases 2 crore sperms which are sufficient for the fertilization of the eggs at the time of laying by the female throughout her life span.

In recent researches in U.S.A. it has been reported that out of 110 queens only 55 mated twice before egg laying. It is also a fact that queen lays fertile and unfertile eggs both in accordance to her will but the factors governing such selective activity are still not known. One queen lays about 1,500 to 2,000 eggs in a day depending upon the seasonal variation and other ecological factors.

The total weight of 100 eggs is equal to her body weight. In the whole life span of two to five years a queen lays about 15,00,000 eggs. When the queen in a colony looses its eggs laying capacity, another worker of the same colony starts feeding on queen's diet *i.e.,* Royal Jelly and develops into a new queen and is provided with the facilities of real queen. At the same time old queen may be driven out but sometimes some workers object that as to why the mother of the colony be driven out so ultimately they also come out with the mother. Sometimes when 2 to 3 queens are developed in a colony, only one takes the position of the real queen and the others come out with some workers to establish new colonies.

Workers: Although the workers are the smallest of the three castes but they function as the main spring of the complicated machinery like honey bee colony. Like the queen, they are also produced from the fertile eggs laid by the queen and live in a chamber called as 'WORKER CELL'. It takes 21 days in the development from the egg to the adult and the total life span of a worker is about 6 weeks.

The workers are atrophid female which sacrifice themselves for the well-being of the colony. The total indoor and outdoor duties of the colony are performed by the workers only. That is why they are provided with some special structures for particular work.

(1) Long proboscis for sucking the nectar.

(2) Strong wings for fanning.

(3) Pollen baskets for the collection of pollen.

(4) Powerful sting to defend the colony against any attack.

(5) Wax gland for wax secretion.

The workers which are engaged in outdoor duties, collect the nectar, pollen, gum and water which are received and stored properly by the house bees. The indoor workers are further sub-grouped for specific duties. Some of them which are very sincere, attend the queen while some others look after the nursery called as NURSERY BEE.

Some produce wax for the formation of the new hive and are known as BUILDER. The repairing of the comb is done by the REPAIRERS. The dead body and other impurities are removed from the hive by the CLEANERS. The fanning in the hive is performed by the wings of the FANNERS. Several other functions like honey storage and ripening are also done by the workers. The guard bee always watches at the gateway. It is said that upto half of the life period workers perform indoor duties and later on become engaged in outdoor duties.

Drone: The drone is the male member of the honey bee colony which fertilizes the queen so called as KING of the colony. They take 24 days to develop from the egg to the adult stage. The sting and the wax glands are absent but in the males the reproductive organs are very well developed. They are reared from an unfertile egg in large DRONE CELL. Drone are totally dependent on the workers and have been seen begging for honey from the workers. The sole duty of the drone is to fertilize the virgin queen. At the time of swarming the drone follows the queen, copulates and dies after copulation.

Introduction of Life

After mating the queen generally lays one egg in one brood cell. The eggs are pinkish coloured, elongated with cylindrical body generally attached to the bottom of the cell. Larvae emerge out from both the fertilized as well as unfertilized eggs. Thus, the larvae from the unfertilized eggs form the drones while the workers are developed from the larvae of the fertilized eggs. Amongst the larvae of the workers one is fed on the royal jelly, a special diet secreted by the young workers in the colony, and becomes the queen of the colony.

The royal jelly consists of digested honey and pollen, mixed with a glandular secretion into the mouth of the workers. After 5 days of feeding the cell is sealed and the larvae undergo pupation. It spins a thin silken cocoon and pupates completely. Emergence of the young ones takes place after three weeks and they get busy in the indoor duties for about 2 to 3 weeks. Later on they are sent for the outdoor duties. All the bees pass through a complete metamorphosis with the various changes in the life-cycle taking place within the comb (Fig.).

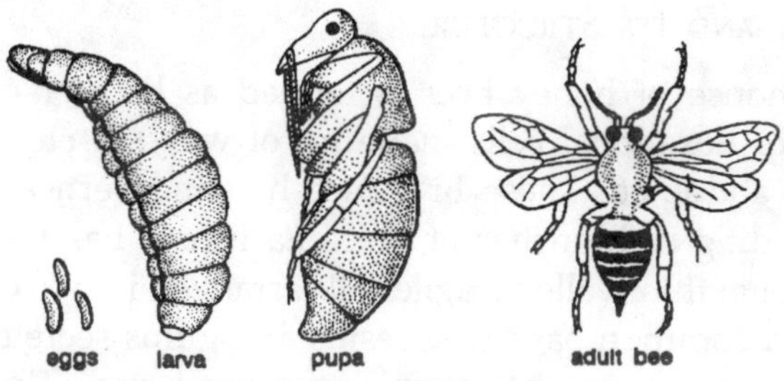

Life history of Apis indica.

Swarming: The process of leaving off the colony by the queen termed as swarming. It happens towards the end of spring or early summer but the real cause of swarming is still not well known. In summers when plenty of food is available and hive is overcrowded by the bees, the queen leaves the hive on a fine fore-noon with some old drones and workers and establishes a new colony at some other place. Now in the old hive a worker is given Royal Jelly and is converted into a new queen of the colony. This new empress of the colony never tolerates her successor, as a natural law in the hive, so she orders to kill the other sisters, if any, in the hive.

Supersedure: When the egg laying capacity of the old queen is lost or it suddenly dies, a new young and vigorous queen takes the position of the old queen and is called as supersedure.

Absconding: The migration of the complete colony from one place to another takes place due to some unfavourable conditions of life, such as destruction of the comb by termites or wax-moths and scarcity of nectar producing flowers around the hive. This phenomenon is quite different from that of swarming.

Nuptial or Marriage Fight: The first swarm is led by the old queen but the second swarm is led by the 7 days old virgin queen which is followed by the drones and is called marriage flight. One of the drones starts copulating with the queen in the sky and fertilizes the queen and dies during the course of copulation. The queen receives spermatophores and stores in the spermatheca. Along with the queen, died drone falls on the ground and the queen reaches the hive.

Bee Hive and its Structure

The house of honey bees is termed as hive or comb. It consists of hexagonal cells made up of wax secreted by the worker's abdomen. These hives are hanging vertically from rock, building or branches of trees. Each hive has thousands of hexagonal thin walled fragile cells arranged in two opposite rows on a common base. The resins and gums secreted from the plants are used for the repairing of the hives. The young stages are generally occupying the lower and central cells in the hive which are the 'BROOD CELLS'. In *A. dorsata* broad cells are similar in shape and size but in other species brood cells are of 3 types *viz.*, WORKER CELL for workers, DRONE CELL for drones and QUEEN CELL for the queen.

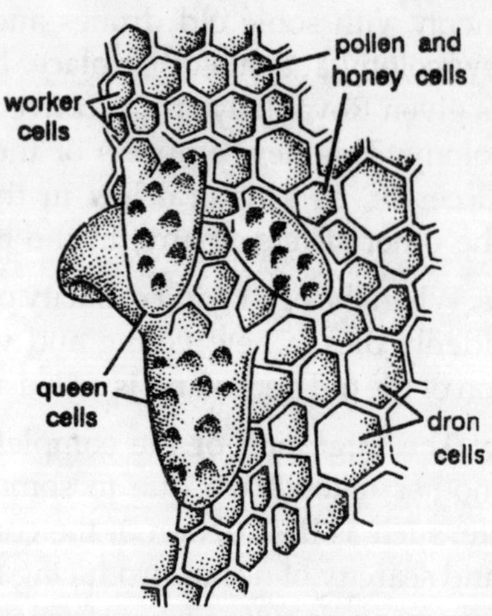

Hive of apis indica showing various cells.

The queen cell can not be used again while the rest are used a number of times. There are no special cells for lodging the adults which generally keep clustering or moving about on the surface of the comb. The cells are mainly intended for the storage of honey and pollen specially in the upper portion of the comb while those in lower part are for brood rearing.

Living Habits

Honey bees are highly organised social insects reported from all over the world. Although they are active throughout the year but in winter season they do little work and do not rear the brood. In spring seasons *i.e.,* at the time of flowering they prepare a strong colony with honey rich combs. They exhibit polymorphism and good division of labour. The bee hives with thousands of individuals are observed hanging down from the branches of the trees and ceilings of houses. The workers communicate informations for the location of the food sources through the 'Waggle Dance', a phenomenon called as 'Language of the bees', by the eminent biologist Karl Von Frish. He has mentioned that the rate of dance is directly proportional to the distance of the food.

Bee Keeping Skills

The ultimate aim of bee keeping is to get more and more honey in pure form. The old method commonly used by old apiculturists is very crude, cruel and of unplanned type. This old method is called as Indigenous method.

Developed Skills

To overcome the drawbacks of indigenous method an advanced method based on scientific facts has been developed. It has opened a new era for the cottage industry in India and has also given an opportunity for lakhs of unemployed persons to keep them busy in this business. From this cottage industry programme the routine agricultural work may not suffer.

First of all care was taken to improve the texture of the hives and during this race hive patterns were introduced in India. The Newton model with 7 to 10 frames (21 × 14.5 cm) in the brood chamber with a shallow super (21 × 6.5 cm sized frames) has been most popular in south, east and central India. Longstroth hive containing 10 frames (44.8 × 23 cm) has been used as a standard hive in Himachal Pradesh, Jammu and Kashmir, and Punjab.

In Uttar Pradesh another type of hive has been in use which was evolved at Jeolikote apiary and contained 8 frames (30 × 18 cm). After gaining experience from the above mentioned hives, Indian Standard Institute has standardised the hives of small and big sizes accommodating frames 21 × 14.5 cm and 31 × 20.4 cm respectively.

Nowadays a typical type of movable hive is constructed which is capable of expansion or contraction according to the requirement of the place, season and climatic conditions.

Appliances for Modern Method

(1) Typical movable hive.

(2) Queen excluder.

(3) Honey extractor.

(4) Uncapping knife.

(5) Other equipments.

Typical Movable Hive: An artificial movable hive is constructed by wooden box based on bee space theory (Fig.). The size and number of frames are variable from hive to hive according to the need. A small space is enough to permit the entrance and exit of workers and drones but queen once placed in hive never comes outside the hive. The perforation size on zinc sheet is only of 0.375 cm but the thorax of the queen is 0.43 cm to 0.45 cm, so the queen can never pass through this pore. This typical hive consists of 6 parts as given below:

(a) *Stand:* It is the basal part of the hive on which the whole hive is constructed. The stands are adjusted to make slope for the hive. Due to this slope rain water comes down quickly.

(b) *Bottom Board:* It is situated above the stand and forms the proper base for the hive having two gates in the front position. One gate functions as an entrance while the other as exit.

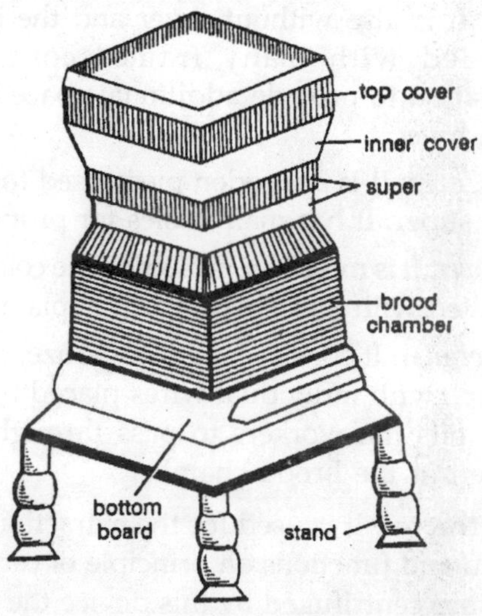

Typical moveable hive.

(c) *Brood Chamber:* The bottom board carries the brood chamber which is the most important part of the bee hive. It is large in size provided with 5 to 10 frames. In each frame a wax sheet bearing hexagonal frames is held up by a couple of wires in a vertical position. Along with the margin of every hexagonal mark, the bees start making wall and ultimately the cells. Here every sheet of the wax is known as COMB FOUNDATION which attracts the bees and provides the base for the comb preparation on both the sides. The frames are kept vertically in brood chamber which is covered over by other frames having a wire meshing through which the workers can easily pass. The comb foundation helps in obtaining a regular strong worker brood cell comb which can be used repeatedly. The Central Bee Research Station at Pune arranged the manufacture of a comb foundation mill which manufactures, different cell sizes required in several regions of the country. The brood chamber is covered by another chamber known as super.

(d) *Super:* It is also without cover and the base. Super is provided with many frames containing comb foundation to provide additional space for expansion of the hive.

(e) *Inner Cover:* It is a wooden piece used for the covering of the super. It has many holes for proper ventilation.

(f) *Top Cover:* It is meant for protecting the colony from rains. It is fitted with zinc sheet which is plain and sloping.

Queen Excluder: It consists of a wire-gauze, extrans guards and drone traps with individual wires placed 0.375 cm apart. It readily permits the workers to pass through it but keeps back the queen in the brood chamber.

Honey Extractor: It is used for the extraction of the honey from the comb and functions on principle of centrifugal force. When combs are centrifuged by this device the pure honey is thrown out without any damage to the comb.

Uncapping Knife: When all of the combs are filled with honey- they are sealed by capping with the wax. So, before such capped combs are placed in the honey extractor, the wax sealing has to be removed with the help of an uncapping knife heated by steam before use.

Other Equipments: Most of the useful equipments for the successful management of the bee are locally manufactured which are very cheap. As they are made locally, they may not be exactly similar to those made at other places. Thus, Indian Standard Institute has standardised some very common equipments for the production of uniform and interchangeable articles. Some materials like protective garments, gum cages, gloves, net veil, bee net, brush etc. are required for easy and well planned handling of the bees.

Advances of Modern Method: In the modern method of bee keeping there are several advantages which encourage the well planned bee keeping.

(1) A proper watch on the activities of the bees can be had.

(2) A strong colony can be developed by providing sugar, syrup, pollen substances to honey bees.

(3) Swarming of bees is checked by modern hive.

(4) The same hive is used again and again so the workers pay their attention more for the honey and not for the hive formation.

(5) Under adverse climatic conditions the hive can be transferred from one place to the other for the protection of the bees.

(6) Comb can be protected from the enemies.

(7) Pure honey in large quantity can be obtained.

Precautions: For the proper management of bee keeping programme following precautions should be taken:

(1) The hive should not be kept more than half a mile away from the place from where the bees have to collect the nectar and the pollen.

(2) People must know about the bee keeper for proper contact.

(3) The boxes must be kept under shade at cool places.

(4) Industry should be near the road for proper transport facilities.

(5) Fresh water reservoir should be near the hive.

(6) Good flora for the collection of pollen and nectar should be there.

Use of Beeswax

Beeswax is a very useful by-product of bee keeping industry. It is yellowish to greyish brown in colour and insoluble in water but completely soluble in ether. Commonly it is a wrong impression to suppose that honey bees convert the pollen into beeswax because beeswax is also a natural secretion of the worker bees and is poured out in thin delicate scales or flakes. Chibnall (1934) has reported that all insect waxes are complex mixture of varying proportions of:

(1) Even numbered alcohols ranging from C_{24} to C_{36}.

(2) Even numbered normal fatty acids from C_{24} to C_{34}, and

(3) Odd numbered normal paraffins ranging from C_{23} to C_{37}.

The various beeswaxes differ only due to change in the proportions of these constituents. Large quantities of beeswax produced and exported, come from *Apis dorsata* bees. Indian Standard Institutions have fixed standards for pure beeswax in order to facilitate its export.

Economic Importance of Beeswax: Beeswax is used in the manufacture of cosmetics, for Catholic churches, face cream, paints, ointments, insulators, plastic works, polishes, carbon paper and many other lubricant. It is also used in the laboratory for microtomy with the common wax for block preparation of tissues.

Hearting Elements of Bees

Enemies of the bees harm the colony in different ways so they have attracted considerable attention in the different regions of the country. The wax moths *(Galleria mellonella and Achroia grisella)*, Wasp *(Vespa* spp. *and Palarus* sp.), black ants *(Componotus compressus)* and bee eaters *(Merops orientalis)* and Kingcrow *(Dicrurus macrocercus)* are common enemies of the honey bee's comb and honey. Man is the last but worst enemy of honey bees.

Before 1958 bees were considered to be free from the diseases though suspected cases of NOSEMA from Punjab and Kashmir were known. But a parasitic *mite-Acarapis woodi* Rennie caused Acarine disease in the adult honey bee in Kulu valley in Punjab in 1956. It was later reported from Himachal Pradesh, Uttar Pradesh and Jammu and Kashmir.

This disease was controlled by the scheme in cooperation with the United States of America at the college of Agriculture Ludhiana, Punjab. Nowadays Indian honey bees are commonly free from any such disease. A strict quarantine measure is being taken to check the spread of any disease from foreign countries. But in European countries bees are commonly attacked by microsporadian which is injurious to bees.

Bee-keeping as Trading

Bee-keeping has gained a good position as an industry in U.S.A., Canada and Australia but in tropical countries it is not growing as per need of the people. Before 1953 attention to bee-keeping was paid only by the State Governments but in the same year, all India Khadi and Village Industries Commission, started to pay attention to it and it was controlled by Union Government itself.

Due to the functioning of the central organisation, bee-keeping industry was spread in South India and in some northern states also. Nowadays beekeeping industry is nation wide and is a good source of cottage industry. Till now limited research and development work has been done in this field which indicates that 'Apiculture' can be raised to the status of a viable occupation in tropical climates provided appropriate scientific and developmental efforts are generated.

Because the tropical countries are very rich in bee fauna, the scientific development of apiculture may be helpful in agriculture and may contribute for providing nutrition, and employment to rural population and may help in raising the economic status of the people.

Advanced Attempts

An international conference on 'Apiculture in Tropical Climates', organised by the 'World Crops' in collaboration with the International Bee Research Association (IBRA) was held at London in 1976. In this Conference it was recommended that the research and development of apiculture should be taken on priority basis in the tropical areas.

The Second International Conference on Apiculture in Tropical climates was organised by the Indian Council of Agricultural Research in Collaboration with the Khadi and Village Industries Commission, Department of Science and Technology and Indian National Science Academy at New Delhi in 1980. In this Conference Eminent National and International scientists and experts realised the importance of apiculture and suggested further research development as a device for raising the industrial status of apiculture.

Indigenous Procedure

Hive: Two types of hives are used in indigenous method of bee keeping *e.g.*, wall or fixed hive and movable hive.

(a) *Wall or Fixed Hive:* It is purely natural type of comb because the bees themselves prepare the hive at any space on the wall or trees. There is an opening on one side through which bees come out of the hive.

(b) *Movable Hive:* It comprises of hollow wood logs, empty boxes and earthen pots etc. placed in varandas of houses. There exist two holes, one is for entrance and the other for exit of the bees. The swarmed bees usually come to the box on their own accord. Some bee keepers use to take the clusters of the swarms from a tree and keep them in the hive.

Extraction of Honey: For honey extraction, burning fire is brought near the bee hive at the night as a result of which bees are either killed or they escape off. Further the hive full of honey is being removed, cut into pieces and squeezed to get honey. Sometimes smoking is done so that the bees may escape from their hives.

Drawbacks of Indigenous Method: The indigenous method of bee keeping suffers from a number of drawbacks due to which it is not recommended by present day panel.

These drawbacks are:

(i) Honey becomes impure because at the time of squeezing, the brood cells, pollen cells, honey cells and larvae are also extracted.

(ii) The colony becomes weak due to killing of the eggs and the larvae at the time of squeezing.

(iii) Formation of new hive by the escaped bees requires extra energy which effects the yield.

(iv) The activities of the bees can be controlled.

(v) The hivation of bees on the same place is only matter of chance.

(vi) The honey robbers, like, rat, ant, wasp and monkeys may affect the hive easily.

(vii) The race improvement programme may not be applied, so no possibility for the selection of the best bee is there.

(viii) The hazards created by climatic factors can not be controlled.

The Products

The chief products of bee keeping industry are:

(i) honey and

(ii) bees wax.

Honey: It is truly an insect product of high nutritive value. The food value of honey may be estimated by the presence of about 80% sugar in it.

Production of Honey: One should not be confused that honey is a direct plant product because the nectar, pollen and cane-sugar bearing secretions of flowers are ingested by honey bees, get mixed with the saliva and undergo certain chemical changes due to enzyme action. At this stage cane-sugar (sucrose) is converted into invert sugars *i.e.,* dextrose and levulose. At this very time some ingredients of bees are also added to the mixture and reduce the water content. The whole mixture is then collected in the honey sac (crop) until the honey reaches the hive. As the honey bee reaches the hive this compound is regurgitated in the hive cell and is known as the honey. Now honey is concentrated by a strong current of air produced by the rapid beating of worker's wings, crawling over the cells.

Honey is very much sweet in taste and white to black in colour with variable smell in accordance with the juices collected from different flowers.

Chemical Composition of Honey: Honey is sugar rich compound having the following constituents:

(1)	Levulose	—	38.9%
(2)	Dextrose	—	21.28%
(3)	Maltose & other sugars	—	8.81%
(4)	Enzymes & pigments	—	2.21%
(5)	Ash	—	1.0%
(6)	Water	—	17.20%

Storage of Honey: After long duration in the stored condition, the honey may be granulated and fermented.

(a) *Granulation of Honey:* The stored honey becomes granular after long duration. Such type of granulation property is the best evidence of pure honey. It is considered that 10 parts of dextrose combine with one part of water, hence forms crystals. Due to less solubility levulose is not crystalised and gives cloudy appearance. Crystallisation is mainly accelerated by the presence of minute air bubbles, colloids and pollens.

(b) *Fermentation of Honey:* After crystallisation honey is subjected to fermentation. Due to crystallisation of dextrose 9% moisture is released, which dilutes the remaining levulose of the honey and the action of yeasts present in air, flowers and soils takes place on levulose and dextrose resulting in the fermentation of honey.

$$C_6H_{12}\dot{O}_6 \rightarrow 2C_2H_5OH + 2CO_2$$

Economic Importance of Honey: Honey is used by human beings in different ways of which the most important is as food and medicine.

(a) *Food Value:* It is estimated that 200 gm of honey provides as much nourishment as 11.5 litre of milk or 1.6 kg cream or 330 gm meat. 2.1 gm of honey provide as much as 67 K. cal of energy. Its sugars, minerals, vitamins and other vital elements are readily absorbed by the systems. Honey may be taken by healthy men as well as those who are ill. It can be taken at any time in any season and by persons of all ages even those just born. It is used in the preparation of candles, cakes and bread. In illness it is preferred over milk because more than half of the body energy is provided burning of dextrose.

(b) *Medicinal Value:* Honey is mildly laxative, antiseptic and sedative, generally used in Ayurvedic and Unani systems of medicine. It is quite helpful in building up of the haemoglobin of the blood and also used as preventive against cough, cold and fever, as blood purifier and as a curative for ulcers on tongue and alimentary canal. Its regular use is recommended after severe cases of heart attack for malnutrition, indigestion and diabetes. It is also found that typhoid germs are killed by honey within 48 hours, those of branchio-pneumonia in 4 days and of dysentery in 50 hours.

(c) *Other Uses:* Other than food and medicine, honey is used in numerous ways. It is used in the preparation of bread, cake and biscuits. It enhances their preserving quality. Much amount of honey goes in making alcoholic drinks. In poultry and fishing industries honey is widely used. In laboratory, honey is used to stimulate the growth of plants, the bacterial culture, inoculation of seeds of cloves, in insect diet and in the preparation of poison baits for fruit flies.

7

ASSESSMENT AND EVALUATION

Once the experimental results are obtained they have to be analysed and interpreted. In experimental situations we may have large number of treatments. Our interest will be to test whether all the treatments have the same effect.

In other words, our intention will be to test the null hypothesis, $H_0 : \mu_1 = \mu_2 = \mu_3 = ... = \mu_t,$ against the alternative hypothesis, $H_1 : \mu_1 \neq \mu_2 \neq \mu_3 \neq ...... \neq \mu_t$. For this purpose we may use Student's t-test by taking all possible combinations. But it is a tedious procedure and theoretically not valid. The appropriate method for such tests is the *analysis of variance.*

The analysis of variance can be regarded as a special development of regression analysis. The fundamental principles of analysis of variance and its most important techniques were developed by R.A. Fisher.

In the early stages, the analysis of variance technique was primarily applied to agricultural and biological experiments. Now it is used in every field of science.

The Principles

There are certain basic assumptions underlying the analysis of variance model.

They are:

1. The effects of the different factors (treatments and environment) are additive.
2. The experimental errors are independent.
3. The experimental errors are distributed normally with mean 0 and common variance σ^2

The usual interpretation of the analysis of variance is valid only when such assumptions are met. When the data deviate from these assumptions to lesser extent the analysis of variance technique will not be affected seriously.

However, larger deviations from these assumptions will affect both the level of significance and the sensitivity of F and t tests. When the level of significance is affected, the results may look more significant than they actually are.

For example, the difference between the effects of two treatments may be significant only at 6 percent level. But we may declare that it is significant at 5 percent level. This will lead to invalid conclusions.

The meaning of the above assumptions, identification of failure of these assumptions and the remedial measures are described in the following sections.

The Process of Adding

When the effects of two factors are independent, the effects of one factor remain constant over all levels of the other factor. In that case, the effects of two factors are said to be additive. For example, consider the hypothetical data in the following table.

Numerical Example for Additivity

Treatment		*Replication*			*Replication effect*		
		I	II	III	II-I	III-I	III-II
1		20	25	35	5	15	10
2		30	35	45	5	15	10
3		40	45	55	5	15	10
Treatment	2-1	10	10	10			
effect	3-1	20	20	20			
	3-2	10	10	10			

From Table, it can be seen that the effect of treatment 2 (treatment 2 -treatment 1) is 10 for ill replications. Similarly, the effect of treatment 3 is 20 (treatment 3 - treatment 1) or 10 (treatment 3 - treatment 2) for all replications. In the same way, it can be found that the effect of replication II is 5 for all the treatments and the effect of replication III is 15 (replication III-replication I) or 10 (replication III - replication II) for all the treatments. Here the effects of treatments and replications are additive.

Let us now consider the hypothetical data in the table below:

Numerical Example for Non-additivity

Treatment		*Replication*			*Replication effect*(%)		
		I	II	III	II-I	III-I	III-II
1		20	30	40	50	100	33
2		30	45	60	50	100	33
3		40	60	80	50	100	33
Treatment	2-1	50	50	50			
effect (%)	3-1	100	100	100			
	3-2	33	33	33			

The percentage increases (effects) are computed as:

$$\frac{30-20}{20} \text{ x } 100, \frac{45-30}{30} \text{ x } 100, \frac{60-40}{40} \text{ x } 100. \text{ and so on.}$$

From Table, it can be observed that the effect of replication II is 50 percent for all treatments. The replication effect is, therefore, not constant over all treatments. The treatment effects are also not constant for all replications. The increase is by a percentage in this case.

When the effects are by certain percentages as in the example, they are said to be *multiplicative* or *non-additive.* If the analysis of variance model for additive effects is given by

$$Y_{ij} = \mu + t_i + r_j + e_{ij}$$

the model for multiplicative effects may be given by

$$Y_{ij} = \mu + t_i + r_j + (tr)_{ij} + e_{ij}$$

For the purpose of examining additivity we may use Tukey's test for non-additivity. For details about this test one can refer Snedecor and Cochran (1967).

In agricultural experiments, the presence of non-additive effects is common in fertilizer trials and insect and disease studies. In fertilizer trials, the plots with low level of available nitrogen in soil, for example, will benefit more from addition of nitrogen than plots with adequate level of available nitrogen. In studies on the incidence of insects or diseases, the incidence pattern will be in multiples of the initial incidence. This happens because of the nature of the growth of insect or disease organisms. In these situations, the assumption of additivity would not be correct.

When non-additive effects are present, the experimental error will be inflated. Consequently, the precision will be lost. In order to overcome this problem the multiplicative effect may be corrected by applying logarithmic transformation.

Analysis of Discrimination

The analysis of variance is the systematic algebraic procedure of decomposing the overall variation in the responses observed in an experiment into different components. Each component is attributed to an identifiable cause or source of variation. The structure of these component parts is determined by the design of experiments.

Suppose that we have a factor *T* with *t* levels (that is, *t* treatments) with equal replication, *r* and their responses, *Y*. The results may be tabulated as shown in the following table.

The Arrangement of Data

	Treatments				
	1	*2*	*i*	*t*	
	Y_{11}	Y_{21}	Y_{i1}	Y_{t1}	
	Y_{12}	Y_{22}	Y_{i2}	Y_{t2}	
	..	..	..	..	
	Y_{1j}	Y_{2j}	Y_{ij}	Y_{tj}	
	..	..	..	..	
	Y_{1r}	Y_{2R}	Y_{ir}	Y_{tr}	
Total:	$Y_{1\cdot}$	$Y_{2.}$	$Y_{i\cdot}$	$Y_{t.}$	$Y..$ = Grand Total
Mean:	Y_1	Y_2	Y_i	Y_t	

The response of the jth individual unit in the ith treatment may be represented by the equation.

$$Y_{ij} = Y_i + e_{ij} = \text{i}^{\text{th}} \text{ treatment mean + random deviation}$$

$$= \mu + (Y_i - \mu) + (Y_{ij} - Y_i)$$

or, $$(Y_{ij} - \mu) = (Y_i - \mu) + (Y_{ij} - Y_i)$$

That is, the variation of an individual response from the

overall mean is the sum of two components. One is the variation in the means of the treatments, Y_i from the overall mean, μ. The other is the variation within each of the treatments. The variation within each treatment is not accounted for by the treatment, and hence it is the error term.

The significance of the difference $(Y_i - \mu)$ is based on the magnitude of the components $(Y_i - \mu)$ compared with $(Y_{ij} - Y_i)$. The statistical test of the significance involves the use of the following identity.

$$\Sigma\Sigma(Y_{ij} - \mu)^2 = \Sigma\Sigma\left[(Y_i - \mu) + (Y_{ij} - Y_i)\right]^2$$

$$= \Sigma\Sigma(Y_i - \mu)^2 + \Sigma\Sigma(Y_{ij} - Y_i)^2 + 2\Sigma\Sigma(Y_i - \mu)(Y_{ij} - Y_i)$$

$$= \Sigma\Sigma(Y_i - \mu)^2 + \Sigma\Sigma(Y_{ij} - Y_i)^2$$

The last term will vanish because $\Sigma(Y_{ij} - Y_i) = 0$

Thus we have,

Total sum of squares = treatment sum of squares + error sum of square,

This can be abbreviated as,

$SS(Y) = SS(T) + SS(E)$

The total variation can thus be partitioned into two components, *i.e.*, variation between treatments and variation within treatments.

Before the magnitude of the two components can be compared, they must be divided by the number of degrees of freedom.

For the component "between treatments", the degrees of freedom is $t - 1$, and that for the "within treatments" is $(tr - t) = (n-t)$ or $t(r - 1)$.

The quantity $SS(T)/t - 1$ is known as the treatment mean square and written as $MS(T)$. The quantity $SS(E)/n - t$ is known as the error mean square and written as *MS(E)*. In general, treatment mean square is denoted by s_t^2 and error mean square by s_e^2. If samples were drawn randomly, s_t^2 and s_e^2 are the independent estimates of the same quantity, σ^2, population variance of *Y*. Therefore, the ratio between two estimates,

$$F = \frac{MS(T)}{MS(E)} = \frac{s_t^2}{s_e^2}$$

follows the *F*-distribution with (t - 1) and (n -t) degrees of freedom. The significance of *F* can be determined in the usual way by using the Tables of *F*. Thus, the *F*-statistic may be used to test hypotheses about the equality of the population means of the treatments, that is, to test $H_0 : \mu_1 = \mu_2 = ... = \mu_t$. If F is significant the treatment means will be significantly different. The analysis of variance concept can be understood by the following examples.

Example

An experiment was conducted to study the effect of four levels of nitrogen on yield of paddy. The four levels of nitrogen were,

No application of nitrogen (t_1),

50 Kg nitrogen per hectare (t_2),

100 Kg nitrogen per hectare (t_3), and

200 Kg nitrogen per hectare (t_4).

The yields of grain in kg per plot are shown in the following table.

Grain Yield of Paddy in Kg/plot

t_1	t_2	t_3	t_4
52	58	62	66
50	52	58	69

Contd..

	t_1	t_2	t_3	t_4	
	56	60	65	70	
	54	57	58	67	Grand
	53	56	61	66	Total
Total	265	283	304	338	1190

Computational steps:

Step 1: Compute the correction factor (CF)

$$CF = \frac{(\text{Grand total})^2}{n} = \frac{Y_{..}^2}{n}$$

$$= \frac{1190^2}{20} = \frac{1416100}{20}$$

$$= 70805$$

Step 2: Compute the total sum of squares

$$\text{Total} \quad SS = \Sigma\left(Y_{ij}^2 - \mu\right)^2 = \Sigma Y_{ij}^2 - \frac{Y_{..}^2}{n}$$

$$= \Sigma Y_{ij}^2 - CF$$

$$= (52)^2 + (50)^2 + ... + (66)^2 - CF$$

$$= (2704 + 2500 + ... + 4356) - CF$$

$$= 71498 - 70805 = 693$$

Step 3: Compute the treatment (or group) sum of squares.

$$\text{Treatment} \quad \text{SS} = \Sigma\left(\text{Y}_\text{i} - \mu\right)^2$$

$$= \Sigma \frac{Y_{i.}^2}{r_i} - \frac{Y_{..}^2}{n}$$

$$= \frac{(265)^2}{5} + \frac{(283)^2}{5} + \frac{(304)^2}{5} + \frac{(338)^2}{5} - CF$$

$$= \frac{1}{5}\left(265^2 + 283^2 + 304^2 + 338^2\right) - CF$$
$$= \frac{1}{5}(356974) - 70805$$
$$= 71394.8 - 70805$$
$$= 589.8$$

Step 4: Compute the error sum of squares (within treatment *SS*).

$$\text{Error} \quad SS = \sum\left(Y_{ij} - Y_i\right)^2$$
$$= \sum Y_{ij}^2 - \frac{\left(\sum Y_i\right)^2}{r_i}$$
$$= \left(52^2 + 50^2 + 56^2 + 54^2 + 53^2\right) - \frac{265^2}{5}$$
$$+\left(58^2 + 52^2 + 60^2 + 57^2 + 56^2\right) - \frac{283^2}{5}$$
$$+\left(62^2 + 58^2 + 65^2 + 58^2 + 61^2\right) - \frac{304^2}{5}$$
$$+\left(66^2 + 69^2 + 70^2 + 67^2 + 66^2\right) - \frac{338^2}{5}$$
$$= \left(14065 - \frac{70225}{5}\right) + \left(16053 - \frac{80089}{5}\right)$$
$$+\left(18518 - \frac{92416}{5}\right) + \left(22862 - \frac{114244}{5}\right)$$
$$= (14065 - 14045) + (16053 - 16017.8)$$
$$+ (18518 - 18483.2) + (22862 - 22848.8)$$
$$= 20.0 + 35.2 + 34.8 + 13.2$$
$$= 103.2$$

Alternatively,

Error *SS* = total *SS* - treatment *SS*

= 693.0 - 589.8 = 103.2

These results can be put in the form of a table. This table is known as analysis of variance (ANOVA) table. For our example, the ANOVA table is as follows:

Analysis of Variance for the Data in Table

Sources of variation	*degrees of freedom (df)*	*sum of squares (SS)*	*Mean squares (MS)*	*F*
Between treatments	(4 - 1 =) 3	589.8	196.60	30.481**
Error	(19-3=) 16	103.2	6.45	-
Total	(20-1=) 19	693.0	-	-

In general, the ANOVA table for one-way classification will be of the form as given in Table.

General form of the analysis of variance table for one-way classification:

Sources	*df*	*SS*	*MS*	*F*
Between treatments	t -1	$SS(T)$	$\frac{SS(T)}{t-1} = s_i^2$	s_t^2 / s_e^2
Error	n - t	$SS(E)$	$\frac{SS(E)}{n-t} = s_e^2$	
Total	n - 1	$SS(Y)$	-	-

For our example, table value of F for (3, 16) *d.f.* and at 1 percent level of significance is 5.29. Since the observed F-value is larger than the table value, it is significant at 1 percent level. This is indicated by two asterisks. If it is significant at 5 percent level it will be indicated by one asterisk.

The columns in the ANOVA table will be same for all types of designs. However, the rows representing the sources of variation will differ according to the type of experimental design. The form of the analysis of variance table and the computational formulae for the sums of squares used in the analysis of variance are given under each design of experiment in the subsequent chapters.

Assessment of Variance Model: We have seen that the jth individual unit in the ith treatment may be represented by the equation:

$$Y_{ij} = \mu + (Y_i - \mu) + (Y_{ij} - Y_i)$$

This equation can be written as

$$Y_{ij} = \mu + t_i + e_{ij}$$

where, $t_i = i^{th}$ treatment effect, and

e_{ij} = error term.

This means that any Y_{ij} is made up of overall mean + treatment effect + an error term. Since the effects are added in this model, it is known as a linear additive model. It is commonly known as *analysis of variance model*. Depending on the design, the analysis of variance model will take different forms. They are given under each design in later chapters.

Analysis of Variance: Its Two Types of Models: Usually the treatments in an experiment are fixed by the experimenter. They are not selected randomly from all possible treatments. For example, in a fertilizer experiment we may select three levels of nitrogen as 60, 80 and 100 Kg per hectare and test whether all these three levels of nitrogen have the same effect on yield of a crop. These three nitrogen levels were not selected randomly from all possible levels of nitrogen. They were determined by us as per our convenience. When the treatments are fixed like this, the analysis of variance model is known as *fixed effects model* or *Model I*.

There are instances in agricultural investigations where the treatments are not fixed like this. For example, we might be interested in the effect of geographic locations on the yield of a crop. It is possible that our concern might be with certain specific locations. In that case we would employ Model I. Instead, if our interest is to generalise to all possible locations from which they have been selected we would employ model II. As another example, suppose that we wish to determine the nitrogen content in paddy plants. We may select 5 hills at random from large number of hills.

From each selected hill three independent determination of the nitrogen content may be made. The results may then be generalised to all hills.

Thus, when t treatments are selected at random from a population of T treatments, we have *random effects model* or *Model II.* It is also known as *variance-components model.*

Under model I, it is assumed that all treatments about which inferences are to be made are included in the experiment. If the experiments were to be replicated, the same set of treatments would be included in each of the replication.

On the other hand, under Model II, a different random sample of t treatments would be included in each of the replication.

For both model I and model II, the equation is same. For example, the equation:

$$Y_{ij} = \mu + t_i + e_{ij}$$

is appropriate for both the models. However, the assumptions underlying these models are not same. In model II, the term t_i is a random variable. The distribution of t_i is assumed to be normal with mean 0 and variance σ_t^2. The variance 'between treatments' will be $\sigma^2 + \sigma_t^2$, where σ^2 is the error variance. The assumptions about the error term are, however, same in both the models.

The procedures will be same for both the models in initial significance tests. After the initial significance test, a Model I analysis of variance is completed by examining the data in greater details. In other words, the null hypothesis under model I is $H_0 : \mu_1 = \mu_2 = ... = \mu_t$. In model II, we will be interested in estimating the variance σ_t^2 rather than the individual t_i. The null hypothesis under model II is, therefore, $H_0 : \sigma_t^2 = 0$. The relative amounts of variation 'between treatments' $(\sigma^2 + \sigma_t^2)$ and 'within treatments' (σ^2) would guide us in designing further studies of this sort. If 'between treatments' variation is significantly more than 'within treatments' (error) variation, we would need more treatments and fewer replications per treatment. On the other hand, if 'between treatments' variation is not significantly more than 'within treatments' variation, we would need fewer treatments and more replication per treatment.

When more number of factors are used, we may have a mixed model. If all the levels of one factor are used while a random sample of the levels of another factor is used in an experiment, a mixed model results.

Model I is better understood than the other models discussed above. Hence, it is used very commonly. It is the most appropriate model for single factor experiments. However, in multi-factor experiments choice should be made between model I and model II.

Homogeneity of Discrimination

In the analysis of variance the usual estimate of error variance is the pooled variance. This pooling is valid only when the variances are homogeneous. That is why the assumption of common error variance is made in the analysis of variance.

The meaning of common error variance can be explained with an example. The error term for individual responses of each treatment can be written as

$$e_{ij} = Y_{ij} - Y_i$$

The mean and variance for these e_{ij} values can be computed for each treatment as given in table given below:

Numerical Example for Common Variance of Error Term

	Treatment			e_{ij} ***for treatment***			e_{ij}^2 ***for treatment***	
1	2	3	1	2	3	1	2	3
2	7	10	-1	1	0	1	1	0
4	8	12	1	2	2	1	4	4
3	4	8	0	-2	-2	0	4	4
1	5	9	-2	-1	-1	4	1	1
5	6	11	2	0	1	4	0	1
Total 15	30	50	0	0	0	10	10	10
Mean 3	6	10	0	0	0	2	2	2

From Table, it can be seen that the mean of error term for each treatment is 0 and the variance is 2.

In practice we may not get the variances exactly equal. However, they should be close to each other. The equality of many variances may be tested by using Barlett's or the test proposed by Hartley.

Hartley's test is simpler as compared to Bartlett's test. When the number of replications is same for all the treatments, the test statistic proposed by Hartley is given by

$$F_{max} = \frac{\text{Largest variance of the } t-\text{treatment variances}}{\text{Smallest variance of } t-\text{treatment variances}}.$$

Hartley has tabulated the sampling distribution of F_{max} statistic. Against the number of treatments *(t)* and the replication degrees of freedom (r - 1), the table value can be read. If the observed F_{max} value is greater than the table value for a specified level of significance, then the variances are declared as heterogeneous.

Yet another simple method for detecting the variance heterogeneity is to use the relationship between the range and the mean. The range tends to be proportional to the variance. Hence the range is used in place of variance. When the ranges and the means of treatments are plotted on a graph sheet, it will result in a scatter diagram.

Usually, heterogeneity of variance occurs when the variance is functionally related to the mean. We have seen some of the functional relationships between the variance and the mean in Section earlier. The functional relationship may also be of the forms, $\sigma^2 = C^2\mu$ and $\sigma^2 = C^2\mu^2$, where C is a constant. These types of variance heterogeneity are usually associated with non-normal distributions.

There is another type of variance heterogeneity caused by the nature of treatments. For example, plots treated with certain insecticides may have higher variation because the insect population is not uniform or due to non-uniform application of insecticides. The variances of varieties that are highly tolerant to moisture stress are expected to be small compared to that of highly susceptible varieties. The heterogeneity of variances may be there due to the genetic variability of the varieties tested. We know that the genetic variability in F_2 generation is much higher than that of the F_1 generation.

Once the variances are found to be heterogeneous, the simplest remedy is to omit the treatments with very large variance from the analysis. If there are too many treatments to be omitted, then this method is not considered satisfactory. Instead, we can use variance stabilising transformations depending on the functional relationship between the variance and the mean. Often the assumptions of homogeneous variance and normality are violated simultaneously. If the variance is not homogeneous then the distribution is not usually normal. Hence the transformations used to normalise the distribution, *viz.*, log, square root and angular transformations, will result in variance homogeneity.

When the variances are not functionally related to the means we may resort to *error partitioning* to handle data with heterogeneous variance. This method should be applied only after eliminating gross errors, if any.

For error partitioning we have to group the treatments with homogeneous variances. For example, suppose that the treatments are paddy varieties and that they are classified as long, medium and short duration varieties having equal variances within each group. Then the sums of squares for varieties and error will be partitioned as follows:

Varieties	Between groups
	Varieties within group 1
	Varieties within group 2
	Varieties within group 3
Error	Replication x between groups
	Replication x group 1
	Replication x group 2
	Replication x group 3

The *F*-values are then calculated as the ratio of the 'group mean square' and the corresponding 'error mean square'. For example,

$$F(\text{group } 1) = \frac{\text{Within group 1 MS}}{\text{Replication x Group 1 MS}}$$

No Punishment against Errors

If the error of an observation is not dependent upon that of another, it is said to be independent. Under certain conditions it is possible that the experimental errors are correlated. In field experiments, it is often found that adjacent plots give similar yields. If we allocate the same treatments to adjacent series of plots the errors tend to vary together.

For example, consider the following layout:

Replication I	*A*	*B*	*C*	*D*	*E*
Replication II	*E*	*A*	*B*	*C*	*D*
Replication III	*D*	*E*	*A*	*B*	*C*

In this layout treatments *A* and *B* are adjacent to each other in all replications. The error for treatments *A* and *B* can be expected to be more related as compared to that between *A* and *D*.

Non-independence of errors tends to produce too many significant results in *F* and *t*-tests.

It is difficult to detect the presence of non-independence of errors by inspection of the data alone. However, by inspecting the layout the non-independence of errors may be detected. The remedy to overcome the lack of independence of errors is the proper randomisation.

Data Transformation: We have seen that in actual experiments the ideal conditions for the analysis of variance may not be obtained. When one or more assumptions fail we are left with two alternatives. One is that a new model can be developed to which the data may conform.

The second is that certain corrections may be made in the data in such a manner that the corrected data meet the assumptions of the analysis of variance model. Development of a new model is not easy. But it is easy to make corrections in the data.

The corrections are done by means of converting the data from their original form to a new form. This conversion of data is known as *transformation* of data. Different types of transformations are available. The most commonly used transformations are log transformation, square root transformation and angular transformation.

Log Transformation: When the original observation Y is converted to log Y, the conversion is known as log transformation. Although log to any base can be used, log to base 10 is generally the easiest. If the observed value is 0, a constant value preferably 1 is added to avoid negative logarithms. When such constant is added, it is added to all the observations.

The log transformation is particularly effective in normalising positively skewed distributions. It is also used to achieve additivity.

An example for log transformation is given in the table below.

Observed values and their log transformed values.

Treatment	*Original Values*			*Log Values*		
	Replication			*Replication*		
	I	*II*	*III*	*I*	*II*	*III*
1	20	30	40	1.30	1.48	1.60
2	30	45	60	1.48	1.66	1.78
3	40	60	80	1.60	1.78	1.90

It can be verified that the treatment and replication effects are not additive in case of original observations. However, the treatment and replication effects are additive after log transformation.

Square Root Transformation: If the original observation Y is converted to a new value by taking its square root, it is known as square root transformation. It is used in case of count data. When some of the observed counts are numerically small, say less than 10, the more appropriate transformation is $\sqrt{Y+0.5}$. The transformation of the type $\sqrt{Y}+\sqrt{Y+1}$ is also used. The square root transformation is used when the observations follow a Poisson distribution.

Angles Transformation: In case of proportions, derived from count data, the observed proportion p can be changed to a new form $sin^{-1}\sqrt{p}$.

This type of transformation is known as angular transformation. It is also known as 'arcsine' transformation or 'inverse sine' transformation.

It may be noted that the angular transformation is not applicable to percentage data which are not derived from count data.

For example, percentage of marks, percentage of profit, percentage of protein in rice, infection index, etc, can not be subjected to angular transformation.

The computational work can be reduced greatly by using ready made table of angular transformation.

In the tables, the values of p are given in percentages.

The angular transformation is not good when

$$p = \frac{o}{n} = 0 \text{ or } p = \frac{n}{n} = 1.$$

The transformation is improved by replacing $\frac{o}{n}$ with $\frac{1}{4n}$ and $\frac{n}{n}$ with $1\text{-}\frac{1}{4n}$, where n is the total number of units under observation.

When all the observed proportions lie between 30% and 70%, the angular transformation need not be used. This is because of the reason that the transformation in that range of values will not alter the conclusions.

The angular transformation is used to normalise the binomial distribution, especially when the observed proportions are in the range of 0 to 30% or 70 to 100%.

All the above transformations are used mainly to stabilise

the variances. For the sake of convenience the situations in which the various transformations are used are summarised below:

Observed distribution	***Functional relationship between variance and mean***	***Transformation***
Poisson	$\sigma^2 = \mu$	Square root: $\sqrt{Y+0.5}$
Empirical	$\sigma^2 = C^2\mu$	Square root
Binomial	$\sigma^2 = \mu(1-\mu)/n$	Angular: $sin^{-1}\sqrt{p}$
Empirical	$\sigma^2 = C^2\mu^2$	Logarithm: Log Y

In making the choice of a transformation, we have three alternative procedures. The nature of data itself may suggest the appropriate transformation. As a second method, the graph showing the relationship between the range and the mean may be used. The third alternative is to use Tukey's test of additivity.

Once the transformation has been made, the analysis is carried out with the transformed data. The conclusions are drawn from such analysis. However, while presenting the results, the means and their standard errors are transformed back into original units. While transforming back into the original units, some corrections have to be made. In case of log transformed data, if the mean value is X, the mean value in the original units will be:

$$Y = \text{antilog}\left[\left(X + 1.15V(X)\right]\right.$$

instead of $Y = \text{antilog } X.$

If the square root transformation had been used, then

$$Y = \left(X + V(X)\right)^2$$

instead of $Y = \overline{X}^2$

In the above formulae $V(X)$ is the variance of the mean X.

Exhibition of Anova Result: We have seen that the null hypothesis of equal treatment means, that is, $\mu_1 = \mu_2 = ... = \mu_t$ can be tested using the analysis of variance technique. The test statistic in this case is *F* which is the ratio of the treatment mean square to the error mean square. The calculated *F* value is then compared with the table value of *F*. The table value is read against the specified level of significance and treatment and error degrees of freedom.

If the calculated value of *F* is greater than the table value, then *F* is significant; otherwise *F* is not significant. If *F* is significant at 5% level of significance, it is indicated by placing one asterisk(*) on the computed *F*-value. If it is significant at 1% level of significance, two asterisks (**) are placed on the computed *F*-value.

A non-significant *F* may result either due to small treatment difference or a very large experimental error or both. It does not mean always that all the treatments have the same effect. When the experimental error is large, it is an indication of the failure of the experiment to detect treatment differences. In order to find out the reliability of the experiment, the coefficient of variation *(CV)* is used. It is computed as

$$CV = \frac{\sqrt{\text{Error } MS}}{\text{Overall mean}} \times 100.$$

If the *CV* is 20% or less, it is an indication of better precision of the experiment. When the *CV* is more than 20%, the experiment may be repeated and efforts made to reduce the experimental error.

In case of significant *F*, the null hypothesis is rejected. Then the problem is to know which of the treatment means are significantly different. Many test procedures are available for this purpose. The most commonly used tests are the *least significant difference (LSD)* and the *Duncan's multiple range test (DMRT)*. The least significant difference is also known as *critical difference (CD)*.

A Great Difference

The CD is a form of *t* test. Its formula is given by

CD = *t* . *SE* (*d*)

where,

$$SE(d) = \sqrt{EMS\left(\frac{1}{r_i} + \frac{1}{r_j}\right)}$$

In case of equal replications $SE\ (d) = \sqrt{\frac{2EMS}{r}}$

In the formula, *t* is the critical (Table) value of *t* for a specified level of significance and error degrees of freedom; and r_i and r_j are the number of

replications for r^{th} and j^{th} treatments, respectively. This formula is derived from the formula for *t* to test the significance of the difference between two means:

$$t = \frac{Y_i - Y_j}{\sqrt{s^2\left(\frac{1}{r_i} + \frac{1}{r_j}\right)}}$$

Two treatments are declared significantly different at a specified level of significance if their difference exceeds the calculated CD value; otherwise they are not significantly different.

The advantage of CD is that it is easy to calculate and it provides a single value for testing all differences. However, there are certain limitations in the use of CD. It should be used only if *F* is significant. It should be used only if the pair comparisons are planned before the data are examined. Comparison of the control with each of the other treatment is a common example for planned pair comparison. If the comparison is selected after the treatment means are observed, a certain number of differences in a set of *t* treatments will be large owing to sampling variation.

Multiple Range Test of Duncan: In a set of *t* treatments if comparison of all possible pairs of treatment means is required we can use Duncan's multiple range test. The DMRT can be used irrespective of whether *F* is significant or not. It involves number of steps which are as follows:

Step 1: Arrange the treatments in decreasing order, that is, according to their ranks.

Step 2: Calculate the standard error of mean as

$$SE(Y) = \sqrt{\frac{s_e^2}{r}} = \sqrt{\frac{EMS}{r}}$$

Step 3: From statistical Tables write the significant studentised ranges (r_p) for $p = 2, 3, \ldots\ t$ treatments and error degrees of freedom.

Step 4: Calculate the shortest significant ranges (R_p), where

$$R_p = r_p \,.\, SE\,(Y)$$

Step 5: From the largest mean subtract the R_p for largest *p*. Declare all the means less than this value as significantly different from the largest mean.

For the remaining treatments whose values are larger than the difference (Largest mean-larges R_p), compare the differences with appropriate R_p value. If two treatments are remaining compare with R_2; if three are remaining compare with R_3; and so on.

Step 6: From the second largest mean subtract the second largest R_p and compare the treatments as in step 5.

Continue this process till all treatments are covered.

Step 7: Present the results by using either the line notation or the alphabet notation to indicate which treatments are on par and which are significantly different.

Example

From an experiment conducted in Randomized Block Design, the following results were obtained: Error mean square = 0.00193; error degrees of freedom = 12; number of replications = 4; mean of treatment 1, $(\bar{T}_1) = 0.36$; $(\bar{T}_2) = 0.16$; $(\bar{T}_3) = 0.21$; $(\bar{T}_4) = 0.16$ and $(\bar{T}_5) = 0.27$.

$$SE(d) = \sqrt{\frac{2EMS}{r}} = \sqrt{\frac{2(0.00193)}{4}} = 0.0311$$

For 12 degrees of freedom and 5 percent level of significance the table value of t is 2.18. Hence,

$CD = t \cdot SE\ (d)$

$= (2.18)\ (0.0311) = 0.0678$

The arrangement of treatments according to their ranks is

$T_1 \qquad T_5 \overline{\quad T_3 \quad T_2 \quad} T_4$

The difference between the means of T_1 and T_5 is 0.36 - 0.27 = 0.09. It is greater than the CD value. Hence, T_1 and T_5 are significantly different. Since other values are less than T_5 value they are also significantly different from T_1.

The difference between the means of T_5 and T_3 is 0.27 - 0.21 = 0.06. It is less than the CD value. Hence, they are not significantly different.

To indicate this a connecting line (bar) is drawn on the top (or bottom) of the treatments. It can be verified that T_5 is significantly different from T_2 and T_4; and T_3, T_2 and T_4 are not significantly different.

The presentation of results as given above is known as *bar chart*. If the treatments are not significantly different they are said to be *on par*.

The DMRT can also be applied for the above data.

Treatment	Mean
T_1	0.36
T_5	0.27
T_3	0.21
T_2	0.16
T_4	0.16

$$SE(Y) = \sqrt{\frac{EMS}{r}} \sqrt{\frac{0.00193}{4}} = 0.0220$$

p	R_p	$R_p = r_p SE(Y)$
2	3.08	0.068
3	3.77	0.083
4	4.20	0.092
5	4.51	0.099

The largest mean - the largest R_p = 0.36 - 0.099 = 0.261. Except the mean of T_5 all others are less than 0.261.

Hence, T_1 is declared as significantly different from T_3, T_2 and T_4. Now, T_1 and T_5 are remaining and their difference is

$T_1 - T_5$= 0.36 - 0.27 = 0.09

Since two treatments are remaining this value has to be compared with R_2. The difference is more than R_2 value (0.068). Hence, T_1 and T_5 are declared significantly different. The second largest mean - the second largest $R_p = T_5 - R_4$

= 0.27- 0.092= 0.178

The means of T_2 and T_4 are less than 0.178. Hence, T_5 is significantly different from T_2 and T_4.

The remaining two treatments are T_5 and T_3. Their difference is 0.27 - 0.21 = 0.06. This is less than R_2 value. Hence they are not significantly different. In other words they are on par. To indicate this a vertical line is drawn connecting them.

The third largest mean-the third largest R_p = 0.21 - 0.083 = 0.127. No treatment mean is less than this value. Three treatments are remaining now.

Hence for comparing the difference between their means, R_3 value (0.083) is taken. It can be seen that $T_3 - T_2$, $T_3 - T_4$ are less than 0.083. Hence all the three are on par.

Instead of using lines alphabets can be used as follows:

T_1	*a*
T_5	*b*
T_3	*bc*
T_2	*c*
T_4	*c*

When two treatments have a letter in common they are said to be on par. If the treatments have different letters they are declared as significantly different.

Class Comparison: In earlier Sub-section we have considered the comparison of pairs of treatments. The pair comparison is used where there are no common characteristics among the treatments.

Sometimes the treatments may have common characteristics in which case the treatments are classified into meaningful groups. Each group will consist of one or more treatments, The aggregate mean of each group is then compared to that of the others.

For this purpose the degrees of freedom and sum of squares for treatments are partitioned into components.

The comparison among *t* treatments has the form

$C_t = w_1T_1 + w_2T_2 + \ldots\ldots\ldots\ldots + w_tT_t,$

where the *w's* are the orthogonal coefficients, *T's* are the totals of the respective treatments, and $\sum w_i = 0$. If C_1 and C_2 are two comparisons such that

$$C_1 = w_{11} T_1 + w_{12} T_2 + w_{13} T_3 + \ldots\ldots\ldots\ldots$$

$$C_2 = w_{21} T_1 + w_{22} T_2 + w_{23} T_3 + \ldots\ldots\ldots\ldots$$

they are said to be *orthogonal comparisons* if the sum of the products of their coefficients is 0. Symbolically,

$$w_{11} w_{21} + w_{12} w_{22} + w_{13} w_{23} + \ldots\ldots\ldots\ldots = 0.$$

If C_t is any comparison among t treatments, the component of the treatments sum of squares is defined by the expression

$$\frac{C_t^2}{r\left(w_1^2 + w_2^2 + \ldots\ldots + w_t^2\right)} = \frac{C_t^2}{D_t}$$

where, r is the number of replications. Any treatment sum of squares having t - 1 degrees of freedom can be divided into t - 1 orthogonal components. This procedure may be explained with an example.

Consider the fertilizer experiment with 6 treatments-2 organic sources of nitrogen, 3 inorganic sources of nitrogen and 1 non- fertilized control.

The comparison of nitrogen sources and control is given by

$$\frac{T_1 + T_2 + T_3 + T_4 + T_5}{5} - T_6$$

where, T's are the totals for the respective treatments. This comparison can be written as

$$1 \cdot T_1 + 1 \cdot T_2 + 1 \cdot T_3 + 1 \cdot T_4 + 1 \cdot T_5 - 5T_6$$

Similarly, the comparison of organic and inorganic sources of nitrogen is given by

$$\frac{T_1 + T_2}{2} - \frac{T_3 + T_4 + T_5}{3}$$

This can be written as

$$3T_1 + 3T_2 - 2T_3 - 2T_4 - 2T_5$$

Observe that the coefficients for one group of treatments is the divisor of the other group. The signs of these coefficients are opposite so that their sum is 0. The above functions are called *linear functions.*

The treatment sum of squares and the degrees of freedom will be partitioned as follows:

Comparison	*df*	*SS*
Control *Vs* Fertilizers	1	$\frac{(T_1+T_2+T_3+T_4+T_5-5T_6)^2}{r\left(1^2+1^2+1^2+1^2+1^2+(-5)^2\right)}$
Organic *Vs* inorganic	1	$\frac{(3T_1+3T_2-2T_3-2T_4-2T_5)^2}{r\left[3^2+3^2+(-2)^2+(-2)^2+(-2)^2\right]}$

The remaining degrees of freedom are partitioned as within organic sources and within inorganic sources.

Trends Comparison: It is often desirable to study the nature of response of the experimental units to the varying levels of treatments like rate of fertilizer application, doses of insecticide and dates of harvest. If the treatments are in equal intervals along an ordered scale like 0, 50, 100 and 150 Kg N/ha, the treatment sum of squares may be partitioned into linear, quadratic, cubic, etc., trend components.

This method of partitioning provides information about the nature of relationship between treatment levels and treatment means. Trend comparison is applicable only when the treatments are quantitative. The orthogonal coefficients for trend comparisons are taken from tables of orthogonal polynomials.

The numerical examples for above types of comparisons are given under different designs of experiments.

Causes of Normality

We know that many statistical tests are valid only when the data follow normal distribution. In some experimental situations the distribution of errors may not be exactly normal as assumed under the analysis of variance model. It may be skewed or it may follow a Poisson distribution or a binomial distribution. There are many methods to test the normality. For testing the normality we need fairly large samples. With small samples it will be difficult to prove normality.

When extreme values are present in a treatment it is an indication of skewed distribution. If the ratio of largest value to smallest value is 2 or more, we may say that there are extreme values.

The nature of the functional relationship between the mean and variance may help in identifying the non-normal distribution. The functional relationship may be $\sigma^2 = \mu$. In such a case, the variance is proportional to the mean. This kind of relationship exists when the distribution is Poisson in form. We can also identify Poisson distribution by the nature of data. Often, counts of events having a small probability of occurrence follow Poisson distribution. For example, count data, such as the number of insects per plant, the number of insects caught in a trap, the number of infested plants per plot, the number of weeds per plot, the number of lesions per leaf, etc., usually will approximate Poisson distribution.

The functional relationship, $\sigma^2 = \mu(1-\mu)/n$ occurs when the distribution is binomial. When the observations are proportions or percentages like germination percentage of seed materials, percentage mortality of insects, percentage of infestation and percentage of barren tillers will approximate binomial distribution. In the usual notation of binomial distribution, $\mu = p$ and $1-\mu = q$.

The non-normal distribution of errors produce too many

significant results. Also, there will be loss of efficiency in the estimate of the treatment means.

In order to normalise a non-normal distribution, we can use data transformation. Logarithmic transformation is used to normalise a skewed distribution; the square root transformation is used to normalise a Poisson distribution; and the angular transformation is used to normalise a binomial distribution.

8

SIGNIFICANCE OF TEMPERATURE

Temperature is one of the most common ecological factor. It has an universal influence and is frequently a limiting factor for the growth and distribution of animals and plants. It also controls reproduction, rate of embryonic development, migration and a number of behavioural characteristics of the organisms.

PROGRESSIVE FACTORS

Since light influences metabolism, it affects the growth and development of organisms. For example, Salmon larvae undergo normal development only when sufficient light is present. In the absence of light their development is not normal and there is heavy mortality. Mytilus lavae in their earlier stages grow larger in darkness than in light.

Effect on Pigmentation: Light induces certain chemical changes which result in the formation of photoreceptors in the form of pigment spots. It may influence the pigmentation of animals in the following ways:

(a) *Skin Colour*: Characteristic lack of pigment in cave

animals is associated with darkness (*i.e.* total absence of light). Certain aquatic animals loose their colour when shielded from light. RASQUIN (1947) showed that cave amphibians (*e.g.* Proteus) and fishes with little or no colour, when exposed to normal light, developed abundant pigment in the skin. The darkly pigmented skins of human inhabitants of the tropics also indicate the same effect.

(b) *Protective Colouration*: The pigmentation of a number of animals acquires a colouration that can protect them from enemies. Such colouration is known as protective colouration. One common type of protective colouration is a simple matching of body colour in respect and pattern to the background. For example, colouration in case of a quail squatting in the grass, moth on the bark of a tree and leaf-insect (Phyllium) among green leaves is exactly like that of the background. It is exceedingly difficult to distinguish them from their surroundings. The arctic hare, weasel and ptarmigan show seasonal colour changes from brown in summer to white in winter. This is clearly related to the conspicuousness of such animals against bare ground or snow-covered landscape.

A second type of protective colouration is obliterative shading in which the birds, mammals or fish display darker colour on the back and lighter colour underneath. This difference counteracts the stronger illumination received from above and the animal blends with its background.

(c) *Colour Changes*: Certain animals have the capability to change their colour according to their surroundings. Frogs and chameleons are well known examples. According to PROSSER and BROWN (1950) colour changes are brought about by visual stimulus. Visual stimulus is the ability of the animals to change the colour to suit the environment in which they live. Colour changes are widely found among crustaceans, insects, fishes, amphibians and reptiles. These help the animals in concealing them from their enemies, help in thermoregulation and are sometimes, associated with breeding (HAMILTON and BART, 1950).

Females of certain birds have dull colouration. This is due to the greater need for congealment while brooding the eggs. The brilliant colouration of males in many animals does not have any protective value for the male itself, but its conspicuousness might draw attention away from the female on the nest. In some birds, such as Wilson's phalarope, the females are brightly coloured and the darb males perform the job of incubating the eggs. The brilliant breeding plumage of the male is often replaced by the duller darb during the winter season. The light is, thus involved in colouration through its effect on reproduction and through its role in protective resemblance.

Effect on Eyes: The degree of development of eyes sometimes depends upon the intensity of light available. In animals living in caves (*e.g.* Proteus anguinus) and in deep sea fishes, the eyes are absent or rudimentary because these animals live in complete darkness. In surface dwelling forms such as crustaceans and fishes *e.g.* Labeo and Catta the ratio of eyes to head is considered to be normal. In the ocean with increasing depth, size of eyes goes on increasing with the progressive decrease in light intensity.

However, below the upper limits of lightless zone, there is a gradual decrease in the size of the eyes. Some deep sea (benthal) fishes have well developed and enlarged eyes to see in bioluminescent light. Terrestrial nocturnal animals such as owls and geckos have large eyes to see in the dark.

Effect on Vision: Higher animals including man are able to see various, objects only in the presence of light. According to BIGELOW WELSH (1924) and CLARKE (1936) many fishes (*e.g.* Lepomis) depend on sight to locate their food.

Effect on Locomotion: In certain lower animals the locomotion is influenced by light. This is known as photokinesis.

For example, the blind larvae of mussel crab, Pinnotheres, move faster when exposed to increased light intensity. Movement of flies is considerably influenced by the wave lengths of light. Locusts stop their flight when the sun is hidden by the clouds.

Phototaxis: In some animals, light plays a role in the orientation of locomotion. This phenomenon of movements of animals in response to light is known as phototaxis. When an animal moves towards the source of light, *e.g.* Rantara and Euglena, it is known as positively phototactic. The animals like earthworms, slugs and certain zooplanktons such as copepods are negatively phototactic as they move away from the source of light.

Phototropism: When only a part of an organism moves in response to light, It is known as phototropism. Phototropism is of common occurrence in plants. Among animals, the hydroids or polyps of many coelenterates and tubicolous worms show phototrophic response.

Photoperiodism: Photoperiodism is the response of an organism to the duration of day-light or length of the day, *i.e.* the time between sunrise and sunset which is known as photoperiod. Between the equator and polar circles, the photoperiod vanes with the season from nearly twenty four hours to nearly no time at all. In temperate regions photoperiod ranges from approximately six to eighteen hours with longer day in the summer and shorter in winter. In equatorial regions the day lasts, for about twelve hours. But it is always the same for a given season and locality. The photoperiod is the most important ecological factor which triggers the physiological; and reproductive behaviour in both plants and animals, such as flowering in certain plants, moulting, fat-deposition, migration and breeding in birds and mammals and the onset of diapause in insects.

Effect of Photoperiod on Plants: In plants a longer photoperiod ensures a greater amount of photosynthesis and subsequent growth. A longer photoperiod leads to an increase in the day temperature and, therefore, increases the metabolic activates of the plants. Therefore, it could be said that in plants there is a delicate balance between metabolic processes in the light and in the dark. The reproductive activities of most of the plants are also correlated with the length of the day. Based upon photoperiodic responses, plants could be classified into :

(i) *Long day Plants:* These bloom and produce seeds in summer on increasing day-length (*i.e.* more than 12 hours).

(ii) *Short day Plants:* These bloom and form seeds when day light period is less than 12 hours. These plants bloom in spring.

(iii) *Day Neutral Plants:* These are those plants whose flowering is not affected by day-length.

Effect of Photoperiod on Animals: Many animals are also directly affected by changes in the photoperiod. The most obvious is the increased reproductive activity of many birds and mammals; migration of birds, mammals and insects; general condition of plumage; food storing behaviour in squirrels; diapause in insects; and number of nitrogen fixing bacteria in legume plants.

(i) *Increased Reproductive Activities:* The increasing day-length increases gonidial development in fishes, birds and mammals. Most of the animals begin to reproduce during spring period so that their young ones could get optimal thermal conditions. In certain fishes longer days are said to induce early maturity (SUNDARAJ, 1972). ROWAN (1921) and WOLFSON (1953, 60) have shown that certain birds migrate towards the south during winter and towards the north during summer. Recently photoperiodism and its effects on domestic poultry and dairy farming are being exploited by poultry keepers and dairy farmers. If hens are kept in artificially lighted pans during winter in order to increase the length of the day, the egg-laying habit of the hens is stimulated. It is believed that light stimulates the pituitary whose secretion in turn stimulates the gonads. Trout which normally breeds in autumn, spawns in summer when day-length is increased artificially in the spring and then decreased in the summer to autumn conditions.

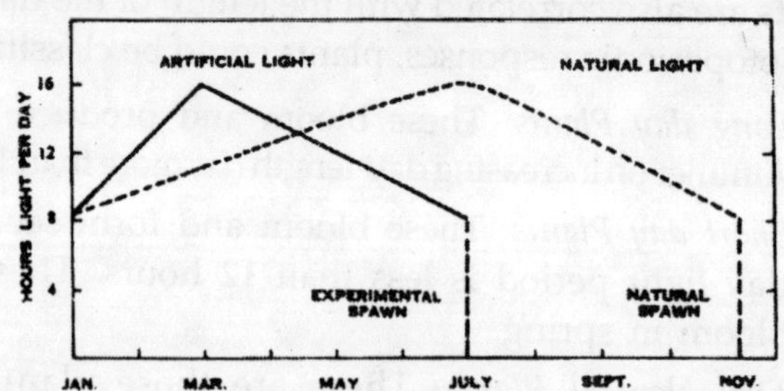

Control of the breeding season of the brook trout by artificial manipulation of the photoperiod (Redrawn from Hazard and Reddy, 1950).

(ii) *Migration:* The movements of migratory animals such as eels, salmons, birds, mammals and even insects are also thought to be triggered by the photoperiod. Each bird has an internal rhythm adjusted to a definite day-length and a bird migrates to such areas where that particular day-length is available.

(iii) *Diapause:* It was WHEELER (1893) who first coined the term diapause. ANDREWARTHA, (1952), defined diapause as a stage in the development of certain animals during which morphological growth and development is suspended or greatly retarded. During diapause in an animal, growth and reproduction stop, metabolic activities are reduced and resistance to extremes of climate develops to the maximum.

Diapause is of common occurrence in insects during winter. It may occur in any stage of development of an insect. The adaptive value of diapause lies in the fact that it enables the insect to preserve itself for more favourable periods and to resist the extremes of climate. The larva of pink cotton bollworm found in southern United States goes into diapause in the months of September and October and remain in this stage during winter. Emergence from diapause begins after spring when the days are slightly loner. But when these larvae were exposed to artificial light in the laboratory, it was found that these larvae could be prevented from going into diapause if this artificial light period ranges from 13-15 hours. In Hieroglyphus nigrorepletus eggs are laid in the soil on bunds and raised grounds during September, October. These eggs remain dormant till the next monsoon in June or July. The hoppers on emerging feed and develop on grass. In the red-hairy caterpillars of groundnut, pupal diapause extends for 9 months in soil. In grasshopper, diapause occurs in the egg stage. The eggs laid down in soil develop upto the blastoderm stage. Thereafter, no development occur. Any further development occurs only if eggs are exposed to low temperatures. Kharpa beetle exhibits low intensity diapause. In this case, when the temperature drops below 30°C, the larvae do not complete growth and development due to lower metabolic rate. But, at the same time they move about, feed a little and sometimes even moult. The diapause is terminated when the temperature rises from 30 to 35°C.

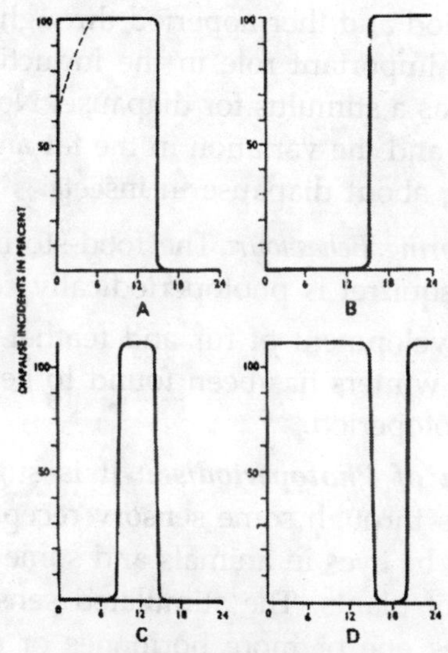

Hours of Light Per Day

Categories of diapause incidence-day length relationship response; among insects.

Photoperiod is an important factor in inducing diapause, since the intensity of diapause is directly related to day-length. In some species the critical day length is within a narrow range of 10-14 hours, while in others the incidence of diapause increase with the increase of day light.

During winter the fall in temperature is associated with a reduction in the number of hours of day light. The diapause is broken by longer day length. This can be illustrated by the diapause of Grupholitha (DICKSON, 1949). Those organisms that entered diapause were found to have hatched during the seasons of the years when there was little more than 12 hours of day light.

A. Long day response;

B. Short day response;

C. Short day-long day response;

D. Long day-short day response.

Long days with high temperatures and short days with low temperatures are associated in the induction of diapause. Thus photoperiod and thermoperiod through their combined action play an important role in the induction of diapause. Food also acts as a stimulus for diapause. Non-availability of a specific food and the variation in the fat and water content may also bring about diapause in insects.

(iv) *Food-storing Behaviour:* The food-storing behaviour of flying squirrel is photoperiodically controlled.

(v) The development of fur and feather coat in animals during winters has been found to be correlated with the photoperiod.

Mechanism of Photoperiodism: It is suggested that the day-length acts through some sensory receptor. The sensory receptor might be eyes in animals and some special pigment in the leaves of plants. The stimulated sensory receptor, in turn, stimulates one or more hormones or enzyme systems that bring about the needed physiological or behaviour response.

Lunar Periodicity: It has been shown that the reproductive cycle of certain different types of organisms show a correlation with the phase of moon. The occurrence of lunar periodicities is well illustrated by the fluctuation in the abundance of conjugants produced by a ciliate living as an ectoparasite on the gills of a freshwater mussel.

The distinct peaks occurred regularly on the days following the new moon. These peaks were not correlated with temperature or other environmental changes. Peaks in the population were seen to occur very regularly on the days following the new moon (RAY and CHAKARVARTHY, 1934).

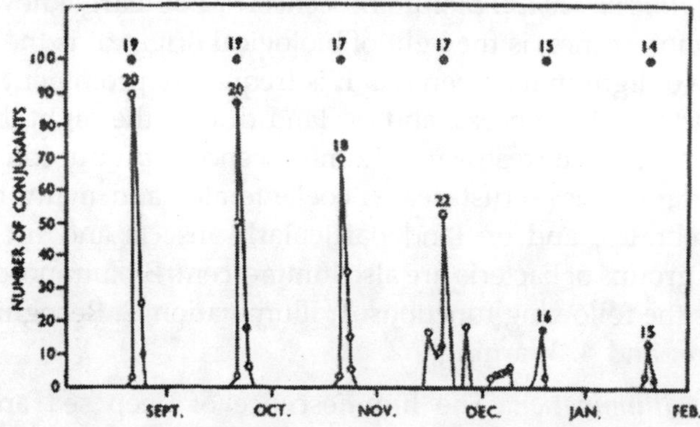

Average daily number of conjugants of the ciliate Conchopnthirius lamellidens on the gills of the freshwater mussel, Laemllidens marginalis. Dates of new moon (top) and of peak numbers are shown.

Most of the organisms exhibiting lunar periodicities are marine. Hence, they are affected by the amplitude of the tide which is greatest at the times of new moon and full moon (spring tides) and smallest at the times of the quarter moon (neap tides). The marine alga, Dictyota, produces its gametes at the time of full-moon spring tide.

The spawning and swarming of a number of marine polychaete worms show a very distinct lunar periodicity coinciding with lunar periods. The palalo worm found in the waters of the South Pacific island comes to the surface in great number on the last quarter of the moon during October and November, producing a luminescence and discharging eggs and sperms into the water.

The worms swim about in small circles in dense masses, giving the sea an appearance of spaghetti soup. In Eunice, swarming occurs during the first and third quarters of the lunar cycle when the intensity of light is low.

The Bermuda fire worm puts on a similar display of fire works in the shallow water early in the evening at the time of full moon. It is due to the drop in light intensity following sunset (HUNTSMAN, 1948).

Bioluminescence: Bioluminescence, or popularly known as phosphorescence, is the light of biological origin. It is the only source of light in the deep sea. It is frequently prominent near the surface of the ocean and on land during the night, but it rarely occures in freshwater. Luminescence is produced in sea by certain fishes, crustaceans, coelenterates and many other invertebrates, and on land particularly insects and fireflies. Many groups of bacteria are also luminescent. Bioluminescence serves the following functions: 1. Illumination; 2. Recognition; 3. Lure; and 4. Warning.

1. *Illumination:* The luminescence of deep sea animal provides illumination for the individuals producing it as well as for other inhabitants. On land and in shallow water, during the night animals may use their photophores as lanterns.
2. *Recognition:* In the deep sea, fishes locate each other by the pattern of lights presented by their luminescent organs. Squids are able to keep together during the dark hours of the night by means of their characteristic flashing. Luminescence is also employed in the recognition of sex, as exemplified by firefly.
3. *Lure:* Animals lure prey by emitting light from their luminescent organs. The predatory fish attract small fish and planktonic invertebrates towards their luminescent organs which are located near the jaws.

4. *Warning:* The sudden flash of luminescent organs may act as a warning to scare off predators. Certain deep-sea shrimps discharge a luminescent secretion into the water. When the cuttlefish is attacked, it discharges its black secretion into the water and escapes from its enemies in the "smoke screen" so produced.

FUNCTIONS OF LIGHT

Light is an important ecological factor which influences the growth and distribution of plants and animals. Light is indispensable for photosynthesis in plants. The life-giving sun's energy is trapped in the carbonic food by green plants. It is responsible for directing and controlling the structural and behavioural characteristics. Light also acts as a limiting factor in the many sided activities of the organisms.

Spectra: The spectral distribution of solar energy that reaches the earth's surface. The solar energy consists of electromagnetic waves composed of high frequency and low frequency wavelengths. In general, high frequency wavelengths from 3600Å (360 millimicrons) downwards are ultraviolet rays, X-rays, gamma rays and cosmic rays and low frequency wavelengths from 7600Å (7.0 millimicrons) upwards are infrared, radar, microwaves and radiowaves.

The part of the solar energy that can be perceived by the human eye is called the visible light. The visible light is made up of a series of colours ranging from violet to red through indigo, blue, green, yellow and orange. The whole range of colours constitutes the visible spectrum. The frequency of wavelengths in visible light ranges from 3600Å 7600Å. In the figure of spectral distribution of the solar energy, about one-half of the total solar radiation is in the infra-region, visible light also constitutes about one-half of the total solar radiation while the ultra-violet light constitutes only a small fraction of the total solar radiation.

Light on Land: The intensity of light reaching the earth's surface depends on two factors-(1) The angle of incidence and (2) the absorption of light rays by the different layers of the atmosphere. The lower the altitude of the sun, the smaller is the angle of incidence but longer is the path of light through the atmosphere with corresponding reduction in intensity. The greatest intensity of sunlight occurs when the sun is overhead. At higher altitudes the intensity of light is correspondingly reduced. The intensity of light is affected by the presence of clouds, moisture and dust in the atmosphere.

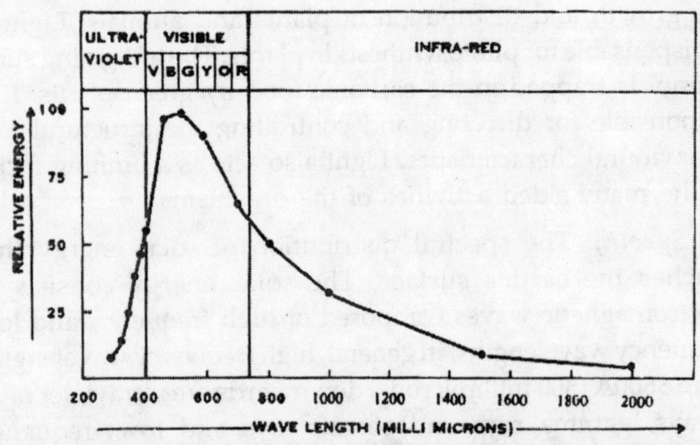

Spectrum of tadiant radiant energy and visible light.

In dense forests, the intensity of light on the forest floor is greatly reduced due to the luxuriant growth of the forest vegetation.

The total amount of light received by an organism is determined by the intensity and duration of light. The duration of light for which an organism is exposed to light is called photoperiodism. At the equator, light duration is of 12 hours. In the temperate regions the length of the day increases with the approach of the spring season, while in the polar regions the day becomes 24 hours long during summer.

Light in Water: About 10% of the total light falling on the water surface is reflected back. The remaining light while passing downward in water is modified in respect of intensity, spectral composition and distribution etc. The suspended particles in water reduce the light intensity both by absorbing and scattering the light.

The rate of reduction in light intensity is known as the extinction rate. Pure water causes the extinction of light at different rates in different parts of the spectrum. Red rays are completely absorbed at a depth of 4 meters.

Orange rays are absorbed at a depth of 20 meters, while the yellow rays penetrate upto 80 meters and green rays upto 80-100 meters. At depth of 100 meters car more the blue light becomes completely predominating. Below a depth of 200 meters or so lies the zone of perpetual darkness.

Percentage of Incident Light

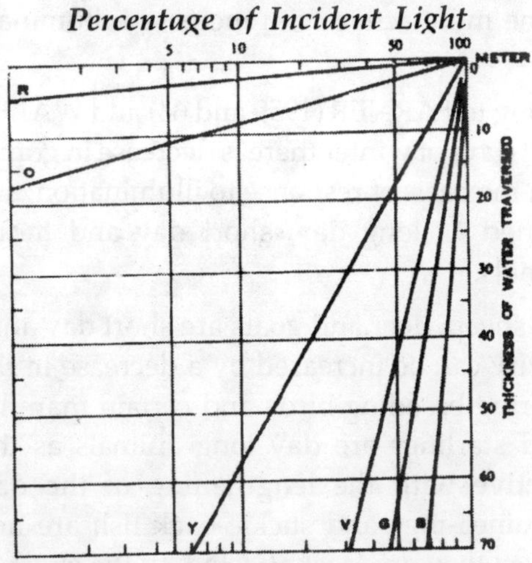

Reduction in the intensity (logarithmic scale) of the colour components of sun light (indicated by initial letters) at increasing deoths (linear scale) in optically pure water.

Biological Effects of Light

Effect on Metabolism: The absorption of light rays of the visible part of spectrum has little effect on organisms. The violet and ultraviolet rays are harmful and produce photochemical changes in the organisms and, therefore, influence their metabolism.

The intensity of solar radiation affects living organisms at different latitudes and seasons of the year. Within optimum limit, photosynthesis increases in direct proportion to the intensity of light.

Cold blooded animals aestivate or hibernate as the solar radiation increases or decreases. Unicellular organisms including bacteria, algae, protozoa and eggs of vertebrates and invertebrates are killed by exposure to ultraviolet rays.

Effect on Reproduction: In many animals light initiates the breeding activities by stimulating theii gonads. There appears to be a definite relationship between length of the day and egg laying in birds. ROWAN (1931) reported that gonads of some birds become more active with increased illumination during summer.

According to FARNER (1959 and 64) and WALFSON (1960) during short days of winter there is decrease in gonidial activity of birds. On the basis of response to illumination, animals have been classified as long day, short day and indifferent day length animals.

Certain sheep, deer and goats are short day animals. Their sexual activity can be increased by a decrease in the length of the day. Spring breeding birds and certain mammals such as turkeys and starlings are day long animals as they become sexually active with the lengthening of the day. Ground squirrels, guinea-pigs and stickle-back fish are indifferent to day length, as they are least affected by the short or long day periods.

Factors Affecting Temperature

The range of temperature varies greatly in different environments such as freshwater bodies, marine and terrestrial. In the open water of ponds and lakes, temperature never falls below 0° and in oceans never falls below-2.5°C. The maximum temperature recorded in marine water has been 36°C in Persian Gulf. On land the minimum temperature of-70°C has been recorded in Siberia and the highest temperature of 60°C in desert areas. The temperature of hot springs may reach upto 100°C.

Variation in Temperature: Generally the temperature of the air near the land surface is 17°C higher in the day time than at night. In deserts this difference may be as much as 40°C. But in water bodies like a deep lake, the difference between day and night temperature is usually less than 1°C. The maximum diurnal change in temperature in marine environment is about 4°C. With increasing depth, the amplitude of day and night change in temperature is reduced and no change is evident below a depth of 15 meters.

Thermal Stratification

***Changes in Temperature*:** Thermal stratification or seasonal changes in temperature can be best studied in large deep lakes of the temperature region. Fresh-water has maximum density at 4°C. Thus any warmer or colder water will float on top of water of this temperature.

During winter, the temperature falls and the surface water is frozen. This ice prevents the cooling of water below it. With the onset of spring the ice melts till temperature rises upto 4°C and entire mass of water has the same uniform temperature.

In summer, the atmospheric temperature may rise to about 25°C. This warms the upper surface layers of water to about 21-12°C. The upper, warm layers being lighter remain at the top while the temperature of bottom layers is hardly 5°C.

In between the two limits lies an intermediate zone with varying temperature from 9°C-21°C. This is the thermocline. This is the zone of rapid vertical temperature change with the thermal gradient of 1°C per meter.

The upper warm layers are referred to as epilimnion and the bottom cold layers are known as hypolimnion. Similar seasonal cycle of temperature occurs in the ocean but it depends on the location and the season.

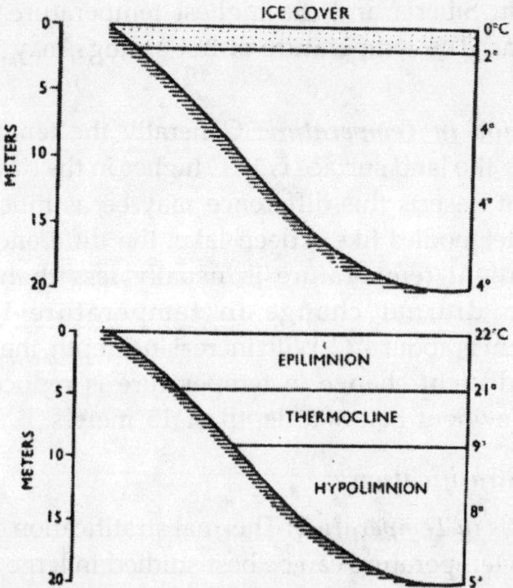

Thermal stratification in a deep lake in temperature region during winter and summer (Redrawn from Concept of Ecology by Kormandy).

The temperature changes in the lakes and ocean are much less and much slower than in the terrestrial environment. Further, organisms can easily get out of excessively high or low temperatures by a short journey into deeper water. Thus life in water is much easier than life on land. But, whenever an unusual change in temperature occurs, dire results may follow with high mortality of the aquatic organisms.

Temperature Tolerance: Every organism has a range of temperature that it can tolerate. Based upon temperature tolerance, organisms are classified as eurythermal and stenothermal.

Animals which can tolerate a wide range of temperature are referred to as eurythermal, *e.g.* cyclops, oysters, toads wall lizard, sperm-whales, man etc. On the other hand animals with narrow range of temperature are referred to as stenothermal, *e.g.* snails, corals, termites, fishes and most reptiles.

The temperature at which the activities of an organism are

at the maximum rate is known as optimum temperature. There are maximum and minimum limits of temperature above and below which organisms of a species could not survive.

Minimum Temperature: The lowest temperature at which an organism can live indefinitely in an active state is termed as the minimum effective temperature. With a further reduction of temperature, the organism goes into a condition of inactiveness called the chill coma. The activities of the animal can be revived by raising the temperature to minimum effective limit. Thus the lowest temperature at which survival is possible is called the minimum survival temperature. The actual value of this temperature depends upon the period of exposure. For example, in laboratory the eggs and larvae of fruit fly, Creatitis capitata, were killed at 7°C on an exposure of 7 weeks, after 3 weeks at 4°C and after 2 weeks at 1°C.

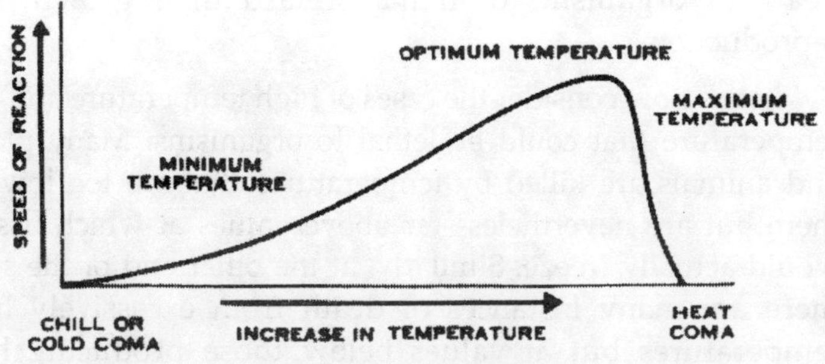

Effect of temperature on the physiological activity of animals.

Maximum Temperature: The highest temperature at which the organism can live indefinitely in an active state is referred to as maximum effective temperature. At higher temperatures the organism goes into heat coma but recovers when restored immediately to cooler conditions.

Generally the range of temperature tolerance within which the living organisms could carry on their life activities lies between 10-48°C. But certain organisms could survive and carry on their metabolic activities at as low as 0°C or as high as 88°C. At low temperature tissue damage and death due to freezing is caused by the formation of ice crystals. At high temperature death occurs on account of thermal denaturation of intracellular enzymes and other cellular proteins.

The favourable temperature range for any particular species is the temperature at which normal physiological activities of the animal takes place. This can be exemplified by the rotifer Keratella procurya (NAYAR, 1971). This species appears in the ponds of pilani, Rajasthan when the temperature is below 24°C and disappears when it rises above 24°C. Its distribution reaches its peak during the months of October to March, when there is a fall in temperature.

Since there is no great difference between maximum and minimum temperatures in water aquatic animals possess a narrow range of temperature tolerance. Exposure to temperatures beyond the favourable range may result in the death of organisms or it may retard their growth and reproduction.

Let us now consider the cases of high temperature and low temperature that could be lethal to organisms. Many plants and animals are killed by temperatures that are too low for them but are nevertheless far above values at which tissues would actually freeze. Similarly, at the other end of the scale there are many instances of death from excessively high temperatures but at values below those producing heat coagulation. Lethal extremes vary greatly from species to species. An organism may be killed by a degree of chilling which is insufficient to cause direct damage to the tissues. As a matter of fact no animal or plant tissues freeze at 0°C. They undergo freeing usually at much lower temperatures. It is a matter of common occurrence that animals and plants die even before their surrounding temperatures have reached 0°C. If the tissues do not freeze, why are these animals and plants killed by moderately low temperatures. This is because a negative correlation exists between low temperatures and chemical reactions.

Most chemical reactions are slowed down by lowering temperature and eventually stop. The stopping of anyone of the chemical reactions or vital process is sufficient to cause the death of the organism.

However, various chemical reactions going in the body come to a stop at different temperatures. Under certain circumstances the organism may even die when all the processes are still going on. It may be due to the fact that the different reactions are slowed down by different amounts with the result that mutually dependent processes get out of tune and cause the death of the organism.

The cause of death of organisms at higher temperature may be due to desiccation and increased rate of metabolism. There is an optimum temperature for every vital process. When the temperature exceeds this optimum, the reactions get out of adjustment and death results.

The rate of chemical reactions increases with the rise in temperature. This has been expressed by Van't Hoff's rule according to which the rate of chemical reactions is doubled for every 10°C rise in temperature. This increase is represented by Q_{10}. In other words Q_{10} represents the effect of temperature on reaction rate, and is in the range of 1.5 to 3 for most metabolic processes. Q_{10} for a particular animal varies with the habitat temperature to which it is adapted.

This lays emphasis on the importance of the choice of temperature range. As a matter of fact the differences in oxygen consumption at comparable stages of development are purely a function of temperature.

For example, during development in the lygaeid bug, the Q_{10} for oxygen consumption is 3-3.8 which drops to 2.5 after the larvae hatch (Glenn Richards, 1964).

Temperature plays an important role in the life of the bed bug (Johnson,1940, 42). The number of generations formed during a year directly depends on the ambient temperature. Thus 12 generations in the tropics, and two in cold climates have been observed. The resulting influence of temperature on the death rate of eggs of Cimex when exposed to three different temperatures for varying periods is shown in figure.

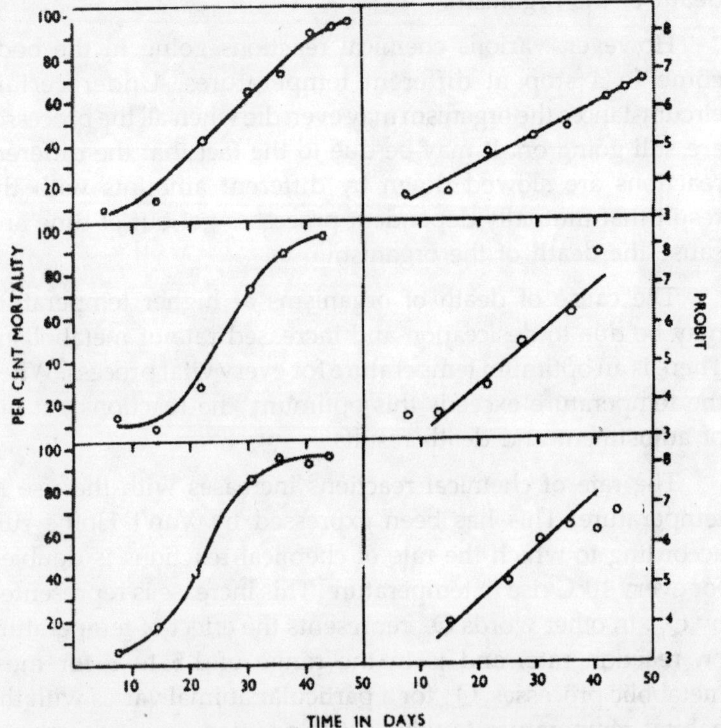

Diagrammatic representation of temperature on the death rate of eggs of Climax. Bottom temperature -42°C, middle -9.8°C, top -11.7°C, Left Death rate percent plotted against time; Right: Probit.

Poikilothermic Organisms: Animals are broadly classified into homoiothennic (warmblooded) and poikilothermic (cold-blooded) animals. In cold-blooded animals body temperature fluctuates with that of the environment, *e.g.* fishes, amphibians and reptiles. The warmblooded animals are those animals which regulate their body temperature at a constant level irrespective of the environmental temperature, *e.g.* birds and mammals.

Meeting Temperature Extremes: The animals and plants have developed certain special mechanisms by which they tend to overcome the harmful effects of temperature. Some of them have been enumerated below:

Formation of Spores, Cysts and Seeds: Certain plants and animals produce spores, cysts, eggs, pupae or seeds that are capable of resisting extremes of temperature. In perennial plants leaves and other aerial parts die off in winter as well as in summer, while their underground rhizomes and stolons lie buried in the soil. On the approach of favourable conditions they give rise to new plants with aerial shoots.

Removal of Water: In some organisms physiological change takes place in the tissues that prevents freezing the bound water by increasing osmotic concentration. This lowers their freezing point considerably. Dried seeds, fruits and cysts avoid freezing by eliminating as much water as possible. Certain dry seeds can germinate even after exposure to 90°C for 3 weeks. Some bacterial cysts can thrive even in boiling water.

Hibernation: Majority of the cold-blooded animals remain dormant during winter season and hide themselves in crevices, under-rocks and in the mud. This dormancy is known as *hibernation*. During hibernation, their metabolic activities are reduced to the minimum possible. At this low rate of activity, little energy is drawn from the body reserves such as fats and glycogen. Thus hibernation is characterised by reduced metabolic rate, low body temperature and reduced rate of heart beat.

Toads and frogs hibernate in solitary burrows on land. Reptiles aggregate in large numbers below rocks or burrows. Aquatic turtles tide over winter buried in mud below the ponds. The common mammals that hibernate are the monotremes, insectivores, bats, some carnivores and rodents. Bats (microchiroptera) are very good examples for hibernators.

Animals like badgers, bears and raccoons also exhibit dormancy in winter. This cannot be called as hibernation as there is no fall in their body temperature during dormancy and this takes place at intervals. This may better be called as pseudohibernation or carnivorian lethargy. The factors that bring about hibernation in animals are the following:

(a) gradually decreasing temperature
(b) lack of food and starvation
(c) inability to move about due to snow covered land
(d) dry food
(e) glandular disturbances
(f) accumulation of fat
(g) lack of proper heat regulating mechanism.

The period of hibernation depends upon the length of the winter. Though some animals hibernate during the whole winter others do so only during the severe months. Hibernation occurs in temperate, arctic and antarctic regions and not in tropical countries.

Aestivation: Aestivation is the period of dormancy that is exhibited by animals during summer. This is commonly met with in insects, some invertebrates and vertebrates. According to Hora some snails remain quiescent during continuous summer rains. The lung fishes spend their summer inside mud cocoons when the ponds go dry and come out only when the rains come.

The Ophiocephalids of the Indo-Malay region remain buried in the rice fields during summer. Terrestrial molluscs like Ariophanta seek cool and shady places during summer and remain dormant by closing the aperture of their shells with a thin membrane, the epiphragm. Among reptiles snakes, lizards and tortoises are good aestivators.

Some Australian frogs become turgid and swollen with water and bury themselves to aestivate till the next rain. Dragon fly nymphs were noticed by Tillyard to aestivate for months. The grand squirrel of California is a good aestivator. The lung fish, *Protopterus,* in order to escape summer extremes, burrows into the mud and secretes a cocoon of slime around itself. It does not feed but subsists on the reserve body fats.

Insects exhibit dormancy in the form of diapause. Diapause is a stage in the life of an organism during which growth and development are suspended or slowed. The groundnut pest, Amsacta albistriga remians dormant (diapause) in the pupal stage for nearly eight months.

Changes during Hibernation and Aestivation: In dormant animals the rate of metabolism is very slow. Though, the frogs loose fat and sugar during hibernation, the development of their gonads proceeds through slowly.

Warm blooded animals temporarily become cold-blooded during hibernation. There is a drop in body temperature in dormant mammals. This may take place in one stroke as in hamster and ground squirrel. Dormant mammals often show heart block when the heart beat rate may be reduced to two or three per minute.

The respiratory movements are slow, and the pulse rate comes down. Hibernating insects can survive lower temperature than what normal insects can tolerate. Dormant animals are usually resistant to parasites. The calcium content of the molluscs is reduced during hibernation.

The activities of the important endocrine glands like pituitary, thyroid and adrenals are retarded. Animals develop special insulating materials like fur, feather and fat during hibernation. Aestivating animals develop a hard covering like cocoon, cyst etc. to protect themselves from drying.

Arousal from hibernation is explosive and certainly the most dramatic aspect of the hibernation cycle. As the animal starts to come out of sleep, its body temperature rises rapidly, perhaps from 4°C to 17.5°C in an hour and half (MAYER, 1960). In ground squirrel, when the anterior part of the body has reached about 36.5°C, the temperature of the hind parts rises rapidly, due to dilation of the blood vessel.

The animal's hair become erect and its body shakes. At 17.5°C the shivering stops and the squirrel moves its tail. At 24°C the animal opens its eyes and suddenly sits up. At this time the temperature rises rapidly again. When the body temperature reaches 24 to 25.4°C, about three hours from the time of arousal, the ground squirrel is warm and active again.

Warm-bloodedness: This is a special device to meet temperature extremes and remain in active condition. Birds and mammals are able to maintain a constant body temperature despite variations in the environmental temperature. In summer the body temperature is brought down to normal by allowing regulated water evaporation from their body surface. Evaporation brings cooling effect to the surface. During cold weather, birds and mammals maintain their normal body temperature by the insulation action of feathers, fur and fat and also by suitable physiological adjustments.

Thermal Migrations: To escape the extremes of temperature certain animals migrate to places where they can have optimum temperature. Such movements are known as thermal migration. Such migrations may be from centimeters to several hundred miles. Thermal migrations include movements from exposed scorching heat to shade in deserts and from shade to sun in cold regions.

Many desert animals have become nocturnal in habit and thus avoid the heat of the day. Frogs, turtles and other amphibians make short trips into or out of water. Burrowing animals escape excessive heat or cold by going deeper into the soil.

With the advent of winter the bear, deer and other game animals descend from the mountains into the valleys. In spring they again return to higher elevation. Many fishes and other aquatic animals leave the shore in summer when the water has become too warm. Insects, birds and mammals undertake long distance migration to escape from climatic extremes. The migration habits of birds and other animals depends not only on temperature but several other factors such as food, breeding and day length.

Effects of Temperature: Temperature has manifold effects on the structure, physiological processes, behaviour and distribution of most of the organisms.

Effects of Metabolism

Temperature has direct effect on the metabolic activity of the organisms. Most of these activities are under the control of enzymes which in turn are influenced by temperature. Initially with the increase in temperature, enzymatic activities also increase, thereby increase the rate of metabolism. For example, the activity of enzyme liver arginase goes on increasing with the increase in temperature from 17°C to 48°C. But any increase in temperature beyond this limit brings about retardation in the activity of the enzymes. Similarly chirping of crickets is higher in warm weather and lower in cool weather.

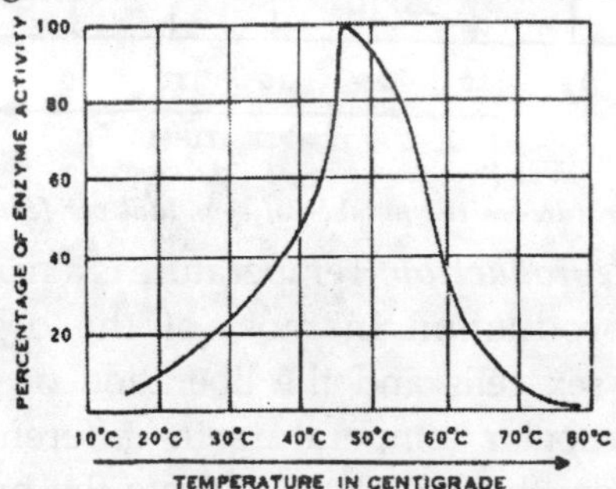

Effect of temperature on the activity of liver enzyme arginase.

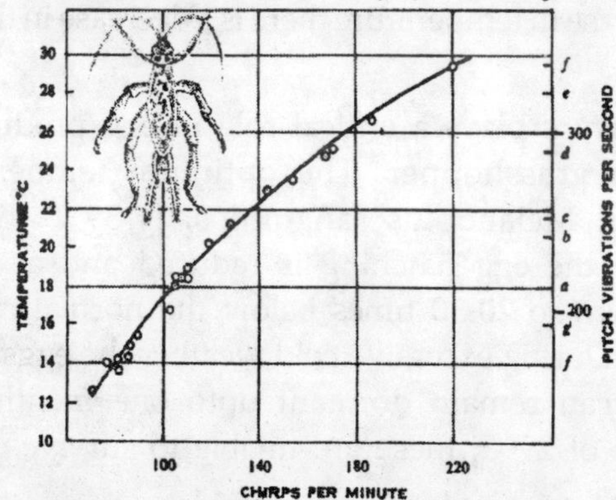

Relation of chirping rate and temperature in tree cricket Oecanthus.

Effect on Reproductive Behaviour: Reproductive behaviour in a large number of animals have been found to be governed by temperature. Some animals breed uniformly throughout the year, some mainly in summer, others exclusively in winter white still others have two breeding periods, one in spring and the other in the fall.

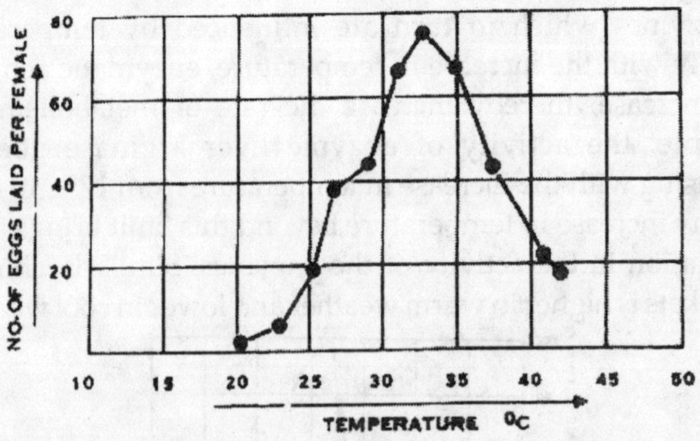

Effect of temperature on the number of eggs laid per female in blow-fly.

Effect on Reproduction: Temperature is a major controlling factor for reproduction in most of the organisms. The maturation of sex cells and the liberation of gametes takes place at a particular temperature in different species. For example, in blow-fly Calliphora sericate the number of eggs laid increases with increase in temperature upto 32.5°C. With further increase in temperature there is a decrease in the number of eggs laid.

Temperature plays a critical role in egg production and hatching in grasshopper. The optimum temperature for grasshopper, Melanoplus sanguinips, is 39°C. Below this temperature the egg hatching is reduced and at 22°C it is brought down to 20-30 times below the normal rate of egg production. During excessive cold weather the eggs freeze. At 23°C these can remain dormant upto one month but at a temperature of 29°C, these are unable to survive.

The egg laying and hatching is also associated with drought.

The grasshopper outbreaks usually coincide with the extended periods of hot and dry weather. But the severe drought also reduces the fecundity of females because of reduction in the succulent food.

Effect on Sex-ratio: In certain animals such as rotifers and daphnids, the sex-ratió is considerably affected by the temperature.

Under normal conditions of temperature daphnids produce parthenogenetic eggs which develop into females. But with the increase in temperature these give rise to sexual eggs which on fertilization may develop into males or females.

But with the increase in temperature these give rise to sexual eggs which on fertilization may develop into males or females.

In the plague flea, Xenopsylla cheopis, males outnumbered females on rats on days when the mean temperature was between 21°C and 25°C. But the position is reversed on more cooler days.

Effect on Growth and Development: Temperature also affects the growth and development considerably at different stages in the life-cycle of an organism.

The optimum temperature for the development of trout eggs is 8°C. The adult trout does not grow in water with a temperature of less than 10°C and the maximum growth takes place between 13-19°C.

In cold, egg development starts at 1°C. It increases regularly upto a temperature of 14°C and above it the egg does not survive. Sea urchin, Echinus esculentus, attains the maximum size in warmer waters.

Mackerel egg starts developing at 10°C. Form 10°C to 21°C, the rate of development increases and at the upper value, the egg will hatch in just 50 hrs. AT 25°C no development takes place. The effect of temperature on the growth of Mackerel eggs can be best understood as shown in the table.

Relation between Temperature and Rate of Development of Mackerel eggs:

S.No.	*Temp.*	*Development*	*S.No.*	*Temp.*	*Development*
1	8°C	No development	5	18°C	Eggs hatch in 70 hrs.
2	10°C	Eggs hatch in 207 hrs.	6	20°C	Eggs hatch in 60 hrs.
3	12°C	Eggs hatch in 150 hrs.	7	21°C	Eggs hatch in 50 hrs.
4	25°C	Eggs hatch in 70 hrs.	8	25°C	No development

(From Elements of Ecology by Clarke)

The time required for the development at various temperatures is also an important feature in the three species of malarial parasite in their:

Species	*16°C*	*17°C*	*18°C*	*20°C*	*24 °C*	*30°C*	*36°C*
1. P. viva	–	halted	–	16 days	9 days	7 days	None developed
2. P. falciparam	–	–	halted	20 days	11 days	9 days	None developed
3. P. malariae	halted	–	–	30 days	21 days	19 days	None

(After Garnthm, 1964)

Bhatia and Kaul (1966) have made elaborate studies on the effect of temperature on the development and oviposition of the red cotton bug Dysdarcus koenegii.

According to them the lower and upper limits for the development of eggs seems to be between 12.5 and 1°C; and 32.5°C and 35°C respectively. With the rise of temperature from 15 to 32.5°C, the duration of incubation period decreases and the viability of the eggs is lower at the extremes of temperature. The nymphs have a narrower range of effective temperature than the eggs (0-35°C) and the duration of the life-cycle at 20, 25, 27.5 and 30°C reduces gradually. With the decrease in the temperature from 35°C to 15°C, there is lengthening of the premating and preoviposition periods.

Effect on Animal Distribution: Extremes of temperature as well as varying period of exposure to heat and cold kill many animals. This is the main reason why temperature acts as a limiting factor and affects the special distribution of animals. Regions of very low and high temperature exclude many species of animals.

Temperature acts as a decisive factor in the distribution of animals in the following ways:

(i) Temperature along with air currents and moisture constitute climate and this influences animals in a composite way.

(ii) Availability of particular types of food in sufficient quantities determines the distribution of many animals and heat affected the distribution of food animals and food plants.

(iii) Thermal stratification brings about thermal zones which vary from two to thirty degree centigrade. The thermoclines between the zones act as barriers for the stenothermal animals.

(iv) Heat alters the density and viscosity of water and thereby reduces buoyancy and this affects the floating ability of many large aquatic animals.

(v) Temperature restricts benthic form to particular coasts. The eastern and western coasts of Africa and India have different benthic animals.

(vi) Animals require for their existence limiting minimal temperature in winter and necessary maximal temperature in summer.

Both these factors together determine the boundaries of distribution.

(vii) The early developmental stages of most of the organisms are vulnerable to temperature changes and this brings about heavy mortality and thereby affect distribution.

(viii) In many animals low temperature prolongs the larval life exposing them to the rigours of life for a long time and thereby the chances of survival are very much reduced.

(ix) A relatively constant temperature or slight variations in temperature always attracts many animals and leads to the concentration of different species.

(x) Rigid stenothermal animals like oysters, lobsters etc. have restricted distribution due to the occurrence of optimal temperature conditions only in some patches in the sea.

(xi) The desert mammals and insects which aestivate during the summer will not reproduce unless they undergo a spell of cooler climate through some part of the year. The gemmules of sponges hatch out after a period of cooling and the eggs of Australian grasshoppers will not grow after a particular stage unless they are exposed to low temperature.

(xi) Bipolar distribution is a standing example of the effect of temperature on distribution.

In general, animals have a three tier distribution according to their living conditions, (A) an area where they have optimum conditions for life, (B) an area within A where they become sexually mature, and (C) an area in B where they can reproduce. The distribution of eurythermal animals is not generally much affected by any variation of temperature.

Temperature and Moisture: The interaction of temperature and moisture depends on the relative as well as the absolute values of each factor. Thus temperature exerts a more severe and limiting effect on organisms when moisture conditions are either very high or low, then when such conditions are moderate. Similarly moisture plays a more critical role in the extremes of temperature when the humidity is low or moderate than when it is very high. Conversely, hot humid weather is less favourable for the weevil.

Structural and Behavioural Effect: Several structural modifications and behaviour patterns are induced by temperature. Animals living in colder regions have a greater longevity than the forms living in warmer zones and are larger in size.

In Drosophila melonogaster, temperature affects the position of genes and chromosomes behaviour during crossing over. This brings about the change in the number of facets in the eyes, size of the vestigial eyes and the presence or absence of more or less eggs than the normal number of six.

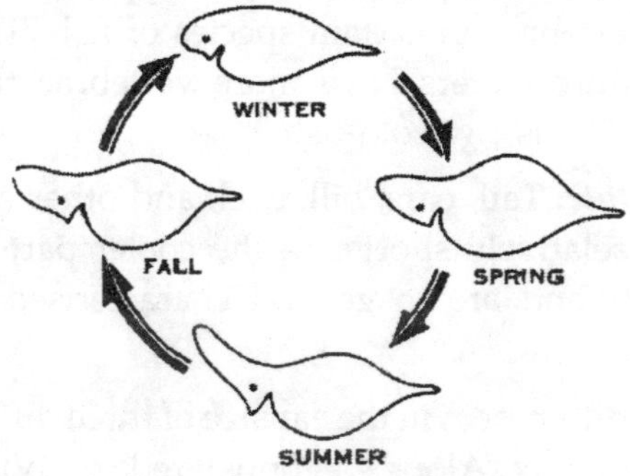

Cyclomorphosis in Daphnia.

Cyclomorphosis: In certain planktonic animals, the body form changes with the seasonal changes in temperature. This phenomenon is known as cyclomorphosis. The process of cyclomorphosis was first described by Coker (1939) in certain cladocerans (*e.g.* Daphnia).

In winter cladocerans have a round head. During spring a helmetlike projection develops on the head. In summer, this projection attains maximum size and in winter the head again becomes rounded. These prolongations of the helmet have been interpreted as an adaptation aiding floatation since the buoyancy of water becomes reduced as the temperature increases (Fig.).

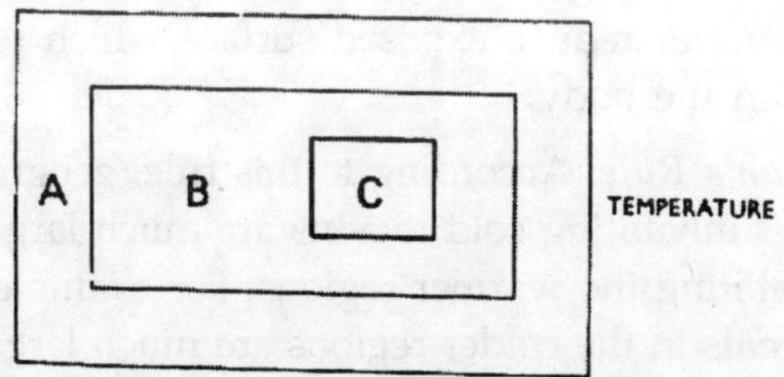

Diagrammatic representation of 3 tier distribution of animals.

Gloger's Rule: Temperature along with moisture and light affects the colouration of some animals. In warm humid climates many mammals, birds and insects are darker in colour than their counterparts living in cool or dry climates.

Jordan's Rule: Temperature has an apparent control on the number of vertebrae in certain species of fish. Thus fishes of low temperature waters have more vertebrae than those of warm-water forms *e.g.* cold-fish.

Allen's Rule: Tail, ears, bill, neck and other extremities of animals are relatively shorter in the cooler parts than in the warmer parts and are in general characterised by compact structure.

It shows differences in the ear size of three different species of fox, an arctic fox (Alopex legopus), red fox (Vulpes vulpes) of temperate region and desert fox (Megalotis zerda).

Eskimos have shorter arms and legs in proportion to their trunks size, which is comparatively larger than in any other contemporary group. Similarly Gazella pincticanda of Himalayas has shorter legs, ears and tail than Gazella benetti found in the plains of Himalayas, though both are of the same body size.

Allen's rule has been tested in laboratory and holds true for species reared in laboratory. It is seen that tails of individuals reared in a temperature range of 15.5°C to 20°C are noticeably shorter than those that are reared in a temperature range of 31°C to 33.5°C, even though these are of the same species.

The adaptive significance of the feature is obvious that short extremities reduce exposed surface which reduces loss of heat from the body.

Bergman's Rule: According to this rule, geographic races of a species inhabiting cold regions are much larger than the forms inhabiting the warmer regions. For example, the birds and mammals in the colder regions are much larger than the related forms inhabiting the warmer regions.

Bergman's rule suggested that as the temperature decreases from the equator towards the poles, the body size of the individuals living there increases gradually. The significance, of the increases in size is explained by presuming that a larger animal has less surface area per unit of weight than the smaller one and thus proportionately less heat is lost by radiation than in a smaller animal. In warmer zones on the other hand the large size will be harmful to these animals, hence they are comparatively short statured. The rule can be supported by facts obtained by the study of nature. The largest bears are the polar bears and the Rodiak bears which are found in the far north while the smaller black bears are found in more temperate regions.

Similarly penguins which are found in Antarctica attain a body length of 1000 to 2000 mm. whereas penguines of the equatorial Galapagos islands are about 490 mm long.

9

INDIAN SCENARIO

The environ scenario of India is very wide indeed. At the first level, special attention must be paid to the school going children and women (about 50% of the population). They are to be made aware of health, nutrition, sanitation, hygiene, development, water and food contamination, fodder and fuel wood etc. Non-Government Organisations (NGO's) have to pay a significant role.

In the Directory of the Dept. of Environment, there are 200 NGO's which work in the area of EE and awareness. Moreover, children are to be told the real meaning of wildlife. They are to be educated for plants, smaller animals, microbes etc. *i.e.* holistic approach to wildlife.

FUNCTIONS OF ENVIRONMENT

Department of Environment could set a plan programme-Environmental Information System (ENVIS) in 1982. It is decentralised system using distributed network of data bases for collection of environmental information.

ENVIS Network with department of environment, its focal

point consists of 10 ENVIS centres on diverse area of environment, established in specialised and reputed institutions in the country.

Objectives of MAB: The general objective of the programme is to develop the basis within the natural and social sciences for the rational use and conservation of the resources of the biosphere and for the improvement of the global relationships between man and the environment; to predict the consequences of today's actions on tomorrow's world and thereby to increase man's ability to manage efficiently the natural resources of the biosphere.

Approach: The programme aims to provide scientific basis to solve real practical problems of resources management through understanding the environmental problems in the ecosystem context.

Organisation: Internationally MAB is guided by International Coordinating Council (ICC) but in India, the MAB is serviced and funded by the Department of Environment, Forest and Wild Life. Indian National Man and Biosphere Committee, was first constituted in 1972 which supervise and directs the programme.

Project Areas: Research under MAB Programme is divided into 14 broad themes which are given.

Project Area 1. Ecological Effects of Increasing Human Activities on Tropical and Subtropical Forest Ecosystems

The broad themes include:

(i) baseline studies in natural tropical forest ecosystem.

(ii) problems involved in the management and natural regeneration of exploited tropical forest ecosystem.

(iii) investigations into the methods of regeneration of exploited tropical forest ecosystem.

(iv) effects of various methods of cultivation and land use on soil structure.

Project Area 2. Ecological Effects of Different Land Uses and Management Practices on Temperate and Mediterranean Forest Landscapes

(i) to identify and assess changes due to man's activities on such landscapes and the effects of these changes on man.

(ii) a comparative study of interrelationship between natural and man made forest ecosystems and impact of changes in human population.

(iii) to develop ways of measuring quantitative and qualitative changes in forest environment.

(iv) ecological impact of forest fire and effect of air pollution on forest ecosystem.

Project Area 3. Impact of Human Activities and Land Use Practices on Grazing Lands

(i) to secure information from natural and social sciences research on grazing lands in order to provide guidelines for management of these lands under different climatic and socio-economic conditions.

Project Area 4. Impact of Human Activities on the Dynamics of Arid and Semi-arid Zone Ecosystems with Particular Attention to Effects of Irrigation

The main aim is:

(i) ecological improvement of arid and semi-arid regions with special emphasis on the effects of irrigation.

Project Area 5. Impact of Human Activities on Mountain Ecosystems

The chief objectives are;

(i) effect of shifting agriculture, over grazing, deforestation

(ii) impact of large scale technology such as hydroelectric power and water storage dams

(iii) effect of tourism and recreation

Project Area 6. Conservation of Natural Areas and the Gene Material They contain.

The main area of interest are:

(i) preparation of list of rare and threatened plant and animal species.

(ii) protection in the form of biosphere reserves and National Parks etc.

(iii) development of techniques for rapid multiplication of threatened and rare plant and animal species.

Project Area 7. Ecological Assessment of Pest Management and Fertiliser use on Terrestrial and Aquatic Ecosystem

The programme lays emphasis on development of methods for control of pests of agriculture and public health importance.

Project Area 8. Ecological Aspects of Urban Systems with Particular Emphasis on Energy Utilisation

The area of research are, energy and material flows into the urban areas and pattern of their utilisation.

Project Area 9. Perception of Environmental Quality

(i) perception of environmental hazards.

(ii) perception of National Parks and other relatively unmodified natural areas.

(iii) perception of quality in urban environment.

Project Area 10. Research on Environmental Pollution and its Effects on the Biosphere

The major areas in the project are:

(i) to identify the environmental indicator.

(ii) acute and long term effects of pollutants from domestic, industrial, agriculture or other sources.

(iii) to monitor the level of different pollutants in the environment.

Project Area 11. Ecological Effects of Human Activities on the Value and Resources of Lakes, Marshes, Rivers, Deltas, Estuaries and Coastal zone 5

The main objectives are:

(i) origin, nature and extent of pollution and its effect on aquatic life.

(ii) sewage problems.

(iii) harnessing the river for agriculture, animal husbandry, drinking and industrial water supply.

(iv) tidal movement effect on ecosystem.

Project Area 12. Ecology and Rational Use of Island Ecosystem.

Project Area 13. Effects of Major Engineering Works on Man and his Environment.

Project Area 14. Interaction between Environment Transformation and the Adaptive, Demographic and Genetic Structure of Human Population.

The above project has following objectives:

(i) Folklore survey and collection and identification of plants and animals used by tribals.

(ii) Conservation of plants used by them.

(iii) Impact of tribal culture on vegetation and wild life.

(iv) Technology of development of tribal communities.

A number of research projects on these aspects have been sponsored.

A unit of Genetic Toxicology is established at Kolkata University to study (i) the chronic effects of sub-toxic doses of pollutant on human systems in relation with different modifying factors as genetic polymorphism, nutrition, etc. and (ii) identify the species of vegetation genetically tolerant to various pollutants.

A project is initiated to study the effect of SO, and particulate matter on vegetation in the urban and industrial areas. The organisations involved are, B.H.U., Varanasi, J.N.U., New Delhi, N.B.R.I., Lucknow, B.S.I., Kolkata National Organisations.

A number of Government and non-government organisations are engaged in environmental studies. Department of Environment Forest and Wild life of India was set up in 1980 to serve as the focal point in the administrative structure of the Government for planning, promotion and coordination of environmental programmes.

Ecological Care

The chief goals of EE in India must be: 1. To improve the quality of environment. 2. To create an awareness among people on environmental protection. 3. To develop the capabilities of decision making.

The spectrum of EE has four major interrelated components *i.e.* (i) awareness, that include making the individual conscious about the physical, social and aesthetic aspects of environment, (ii) real-life situations, that link environment to life, these conditions are locational specific, thus problems and properties of each area may be different, (iii) conservation and (iv) sustainable development, where spot light would be on utilisation and not on exploitation. In the former, the idea is that all resources are finite and there is also a limit to the growth of living systems.

Thus resources are to be used in wise manner. Sustainable development aims at utilisation of resources not only by the present but also by the future generations in a manner that utilisation (and not exploitation) is balanced. Utilisation of resources for development is always associated with some negative impacts. Thus efforts are to be made to contain or minimise them.

Primary School Stage: The attempt is made to sensitise the child about environs. Emphasis should be mostly (75%) on building up awareness, followed by real life situation (20%) and conservation (5%). The contents to be used are surrounding from home to school to outdoor situation. Teaching strategy includes audio-visual and field visits.

Lower Secondary Stage: At this level objective must be real life experience, awareness and problem identification. The quantum of awareness must decrease with increase in real life situations. The contents are supplemented with general science. Teaching, practicals and field visits are to be made.

Higher Secondary School Stage: The emphasis must be on conservation, assimilation of knowledge, problem identification and action skills. Contents may be science based and action oriented work. There should be proper teaching, practicals and field work.

College Stage: Maximum emphasis should be on knowledge regarding sustainable development based on experience with conservation. The contents must be college based on Science and Technology. Teaching practicals and action-oriented field work is to be done. In the school education NCERT has been playing vital role in designing syllabi, text books, guide books, charts, kits, teaching materials and other aids.

University Education: EE at this level is being looked after the UGC. There are about 10 universities teaching environmental sciences. The University education has three major components- Teaching, Research and Extension. At post graduate level, four major areas are recognised:

1. *Environment Engineering:* It includes subjects like architecture, civil engineering, town and country planning, including human settlement, slum improvement, landscape architecture, industrial design, regional science and urban ecosystem studies.
2. *Conservation and Management:* It includes fields like land use, forestry, agriculture, energy, waste management, wild life management, national parks,

biosphere reserves, biological diversities, water management, mining management, non-polluting renewable energy development etc.

3. *Environmental Health:* This deals with public health and hygiene, sanitary and chemical engineering, occupational health, toxicology, nutrition and drug use etc.
4. *Social Ecology:* It includes subjects like ecology, sociology, social planning, cost-benefit, community organisation and services, psychology and counselling, environmental ethics and related areas of humanities.

There are some institutes, centres assisted by Dept. of Environment which provide formal education/training in environmental areas *e.g.*, Centre for Environmental Education (CEE), Ahmedabad, Indian Institute of Forest Management, Bhopal and Indira Gandhi National Forest Academy, Dehradun. This education is designed for any age group, participating in cultural, social, economic development of the country. They form clubs and arrange exhibitions, public lectures, meetings, environmental campaigns. Following are the main constitutes of this education.

Adult Education: Adults may influence the society to protect the precious environs by generating posters, slides, audio-visual and information pictures.

Rural Youth and Non-student Youth: They may act as volunteers.

Tribals and Forest Dwellers: They are an important media to protect the forest wealth.

Children Activities: Dept. of Environ with the help of United School Organisations of India organised essay competitions among different age group children. School term courses are also given by NMNH in EE every year. The National Museum of Natural History (NMNH) conducts spot painting, modelling and poster design about environment for children.

Eco-development Camps: A set of guidelines has been prepared by D.O. En (1984). The objectives are: (i) To create awareness in youth about basic ecological principles. (ii) To identify root cause of ecological problems as related to human activities. (iii) To promote for solving ecological problems. (iv) To develop a spirit of national integration.

Non-Government Organisations: There are more than 200 NGOs of which most are involved in EE and awareness. Others in pollution control, nature protection and conversation, rural development, waste utilisation, wild-life conservation, floristic and funal studies, afforestation and social forestry and eco-development.

Public Representative: India has environmental forums for MPs and MLAs to discuss environmental problems facing the country. They may be sound public opinion and stimulate public interest for saving the environs.

Training Executives: Regular courses should be arranged for environ activities among administrators.

Research and Development Programme: Such R&D efforts are supported by D.O. Environment in Biosphere and Man as well as solving basic and applied environmental problems.

Foundation Courses: The courses for the probationers selected for IAS, IFS, IPS and cadets of three wings of Armed Forces need to be supplemented with foundation courses on environment relevant to their area of specialisation.

Development of Educational Material and Teaching Aids: Materials for media (T.V., radio films, newspaper etc.), audio, mobile exhibitions, audio-visual materials must be operated by competent manpower. One such centre in India is centre for Environmental Education, Ahmedabad.

Development of Trained Manpower: Dept. of Environment (DOE) must organise training programmes for the professors, technical personnel, lecturers and legal experts.

National Environment Awareness Campaign or National Environment Month

Commencing from 1986, DOEn conducts NEAC and NEM, from November 19th to December 18th every year is observed as NEM. Each year there is major environmental theme.

World Environmental Day

All Govt. in the states, UTs, Universities, Schools, Colleges, academic institutions and voluntary organisations organise suitable activities on WED, *i.e.* 5th June of each year DOE supports the function financially.

National Environment Awareness Campaign/National, Environment Month

Commencing from 1986, Department of Environment conducts NEAC and NEM; from November 19th to December 18th every year is observed as NEM. Each year there is a major environmental theme.

Study Material

The environmental education may be thought not as an end in itself but as a means to an end. Education is the process of development of child's personality.

The Review Committee (1977) on curriculum of EE has emphasised the need for stressing more environment based education. The committee has also recommended that physical and bio-sciences, life sciences curriculum should be made environmental oriented with emphasis on problem related to environment.

The review of literature on the theme of environment reveals that it is related to physical, biological and social sciences, thus, the edifice of environment, knowledge and understanding is interdisciplinary in nature. Therefore, it can be adequately designed and included in the course curriculum of teacher education programme.

The teaching methodology of physical sciences, bio-science and social sciences are taught and practised under the main course of teacher-education.

Therefore, the related concept and components of environment can be easily included in these teaching subjects of teacher education curriculum.

The universities and educational institutions have designed the course content of Environmental education for B.Ed, and M.Ed, level. The main emphasis has been given on physical and biological component of the environment.

Tiresome Courses

1. *Historical Development of E.E.:* Ancient view point and recent concept of E.E.
2. *Concept of Environment:* Meaning structure and processes. Types of environment and components.
3. *Concept of Education:* Meaning, objectives, content and status, interdisciplinary approach to education.
4. *Concept of Environmental Education:* Relationship between Environment and Education, meaning and definition characteristics and objectives, content, method and techniques.
5. *Concepts related to E.E.:* Ecology, Ecosystem, Autocology, Biomes and Pollution.
6. *Physical Environment and Education:* Meaning components, Air, Water, Sound/Noise Pollution. Meaning, structure, components, factors affecting environmental, industry, urbanisation, traffic , etc. and role of education and interdependency of various factors of environment.
7. *Social and Cultural Environment and Education:* Meaning and structure, Pollution, colonisation, urbanisation due to social changes and mobility. Development of media, factors influencing the environment and role of teachers and schools, method health and community health.

8. Agencies of Environmental Protection and Information-Steps taken. The governmental organisation, private organisation, legal steps, role of educational institutions.

Teaching Techniques

As already mentioned that the contents of EE is largely interdisciplinary in nature. It is both art (doing) and science (understanding), organised from primary to university level, the objectives of EE are not confined upto knowledge and awareness but include skills, attitudes and values.

Thus strategies of teaching and learning have wide coverage, based on content, its components, levels of education and the objective of E.E.

The major objectives of E.E. are awareness, attitude and action. Therefore, a student should be allowed and asked to observe simple phenomena of earth and sky. The teacher should also translate awareness in his action which will be the model for the students. Thus the doing part much more important for the teacher.

The role of teacher is very significant in realizing the above objectives of E.E. A number of projects can be assigned to the students in school situations for making herbarium, plantation in school campus. It is the teacher who can sensitise his students for improving the quality of environment.

From the given table it is evident that E.E. employs scientific and non-scientific methods of teaching. All the methods provide the awareness but skills and attitudes are equally important objective in teaching environment education.

Strategies of Teaching of Environmental Education with Reference to Objectives and Levels.

Strategies of Teaching	Objective of Environmental Education			
	Awareness	Skills	Attitudes	Action
Primary Level				
1. Observation method	+ + + +	———	++———	———
2. Playway method	+ + + +	+——	+ +———	
3. Field trips	+ + + +	+———	++———	———
4. Dramatisation	+ + + +	———	++———	+———
Secondary Level				
1. Lecture method	+ + + +	———	++———	———
2. Questions-Answer	+ + + +	———	———	———
3. Project method	+ + ——	———	———	———
4. Educational tours	+ + + +	———	++———	———
5. Dramatisation	+ + + +	———	++———	++———
6. Observation	+ + + +	———	+++———	———
Higher Level				
1. Lecture method	+ + + +	———	++———	———
2. Group discussion	+ + + +	———	++++	———
3. Seminar and w.s.	+ + + +	+++———	+++———	++———
4. Survey method	+ + + +	———	+++———	———
5. Action research	+ + + +	+++———	+++———	++++——
6. Experimentation	+ + + +	++———	+++———	++———
7. Interdisciplinary approach.	+ + + +	———	++++———	++———
8. Demonstration	+ + + +	++———	++++———	++———

The objectives of E.E. can be realised through formal and non-formal system of education. The strategies of E.E. have been summarised in the above table:

Practical Application: Ryan developed 'Educational Excursion Method' of teaching which is commonly used in geography, botany, zoology, study of nature and history teaching.

There are a number of topics in E.E. which can be taught effectively by this method.

TECHNOLOGY AND ITS BENEFITS

The following are the theoretical basis of this method.

(i) It provides the real experience about the E.E. to the learners.

(ii) It involves the observations imagination and ability of discovering the cause effect relationship among the environment components.

(iii) It develops the feeling of cooperation and group work as social principles.

Objectives

(i) To develop the tendency of excursion and its utility for the understanding.

(ii) To provide awareness about the environment.

(iii) To develop the ability of observation, imagination and discovering.

(iv) To develop ability of cooperation and team work.

The following steps are taken is organising educational excursion.

(i) The specific objectives are to be formulated.

(ii) A schedule—date, time, number of students, duration and financial aspects should be finalised.

(iii) A guide sheet is prepared and supplied to the students.

(iv) The students are asked to prepare a report about their observations and visiting different places.

(v) The visiting organisations are to be contacted for the arrangement.

(vi) The excursion should be followed group discussion to highlight the environmental components and source of pollution.

Uses

(i) Teaching can be made effective and interesting by real experiences.

(ii) It helps in developing ability of observation and imagination.

(iii) The objectives of teaching can be realised by this method.

(iv) It develops feeling of cooperation and team work.

The co-curricular activities of education are the most appropriate means for providing opportunities for the action. The teacher is the main means for implementing the programmes and realizing and objectives of E.E. and organisation of co-curricular activities in schools and outside the school.

There are two main programmes recommended by Educational Commission and policy *i.e.* National Social Scheme (NSS) and socially useful productive work (SUPW). The activities organised through these programmes are:

(i) To clean the environment through NSS camps.

(ii) To grow plants and develop gardens for protection.

(iii) To clean the public places and parks etc.

(iv) To construct roads and dig pits for the wastes.

(v) To develop sense of sanitation among the people by organising cultural programmes.

(vi) To develop the consciousness about population education by organising camps of population education or family planning.

(vii) To encourage the students to prepare charts related to environmental pollution and its protection.

(viii) To motivate the students for using stories and essays on the environmental education.

The Review Committee (1977) on curriculum E.E. has stressed the need for more environment based education. There are more than two hundred non-government organisations engaged in environmental education awareness and training youths on different areas.

Recently the university departments are organising seminars and workshops on E.E. to develop the course content for B.Ed and M.Ed classes.

The courses are to be designed at two levels - first in the core course and second is the specialised course. The core course is compulsory for B.Ed students and specialised course is the part of teaching methodology course.

Different Programmes

1. Aswan High Dam in Egypt (1964). It is the biggest dam in the world. It was constructed at Aswan in 1964 on river nile. It has caused several environmental problems of serious consequences because the likely or probable impact of dam and reservoir on environmental conditions were not properly assessed. It caused several kinds of biotic and abiotic problems.
2. Earthscan. An agency founded by UNEP in 1976 that commission's original articles on environmental matters and sells them as features to newspapers and magazines especially in developing countries.
3. Convention on International Trade in Endangered Species. (CITIES). An international forum whose membership for agreement is open to all countries. For India, the Ministry of Environment and Forests functions as nodal agency for participation in international agreements.

4. Environmental Protection Agency (EPA). This is an independent Federal Agency of the U.S. Government established in 1970. It deals with protection of environment by air, water, solid wastes, radiations, pesticides, noise etc.
5. European Economic Community (EEC). It is a community of 12 European nations with sound political, economic and legal base. The community has joint agricultural and scientific programme between its members. It also assists other countries in environmental programmes. It has programmes of framing and implementation of coordinated policy for environmental improvement and conservation of natural resources. CPCB, India has recently taken up a project on air quality monitoring with assistance of EEC.
6. Human Exposure Assessment Location (HEAL). The project is a part of the Health Related Monitoring Programme by WHO is cooperation with UNEP. The project has three components *viz.* (i) air monitoring (ii) water quality monitoring and (iii) food contamination monitoring on a global basis.
7. International Council of Scientific Unions (ICSU). A non-governmental organisation, based in Paris, that encourages the exchange of scientific informations, initiates programmes requiring international scientific cooperation and studies and reports on matters to social and political responsibilities in treatment of scientific community.
8. International Union for Conservation of Nature and Natural Resources (IUCN). An autonomous body founded in 1948, with its Head quarters at Morges, Switzerland, that initiates and promotes scientifically based conservation measures. It also cooperates with United Nations and other inter-governmental agencies and with sister bodies of world wildlife fund (WWF).

9. Trans-Alska pipe line in USA (1977). This project was initiated to carry mineral oil and natural gas through 1270 kms long underground pipelines from the newly discovered oil wells near prudhoe bay on north Coast of Alaska to port Valdez to south coast of Alaska. This project has not affected Alaska and marine and environment except the political problems that the route and pipe line involves Canada. It is not advisable due to safety and strategic point of view.
10. International Marine Consultative Organisation (IMCO). It regulates the operation of ships in high seas, from marine water pollution viewpoint.
11. United Nations Educational, Scientific and Cultural Organisation (UNESCO). A United Nations agency founded in 1945 to support and implement the efforts of member states to promote education, scientific research and information and the arts to develop the cultural aspects of world relations. It also holds conferences and seminars, promotes research and exchange of informations and provides technical support. Its Headquarters are in Paris. Independently as well as in collaboration with other agencies like UNEP, it supports activities related to environmental quality, human settlements, training to environmental engineers and other socio-cultural programmes related to environment.
12. United Nations Environment Programme (UNEP). An UN agency, co ordination of inter government measures for environmental monitoring and protection. It was set up in 1972. There is a Voluntary United Nations Environment fund to finance environmental projects. There is an Environmental Coordination Board to coordinate the UNEP programmes. Its Headquarters are in Nairobi, Kenya. UNEP was founded to study and formulate International guidelines for management of the environment. UNEP is assisting many such programmes in India.

13. Earth watch Programme. A worldwide programme, established in 1972 under the terms of the declaration on the Human Environment. It monitors trends in the environment based on a series of monitoring stations. Its activities are coordinated by UNEP.
14. Man and the Biosphere Programme (MAB). The programme is the outcome of International Biological Programme (IBP) that has already concluded its activities. MAB was formally launched by UNESCO in 1971. There are 14 project areas under this programme. We shall provide here the details of MAB with special reference to major activities in our own country so far done under the same, and the priority areas for future.
15. Project Earth. Developed in collaboration with UNEP to inspire, interest and educate young people world wide on crucial issues facing the Earth's Environment. The project is led by Mr. Robert Swan, UNEP Good will Ambassador for youth. He is the only person to have reached the North Pole and the South Pole on foot.
16. Tehri High Dam Project (India). It is one of the most controversial river projects in world as regards the conflict between environmentalists, local people, project authorities and government. It is being constructed on the Ganga river below the confluence of its two main tributaries-Bhagirathi and Bhilangane at Tehri of Uttar Pradesh. Tehri Hydro corporation (THDC) was formed in 1989 and construction work was taken over by Union Government of India with soviet technical and economic aid. It is 260.5 meter highest rock fill dam in the country. Before planning this dam, proper assessment was not done to visualise the probable adverse effects.
17. Sardar Sarovar (SS). Project near Navagam in Bharauch district of Gujarat is one of the costliest projects affecting villages in three states - M.P., Maharashtra and Gujarat. About 245 villages will be submerged of which about

193 in M.P. alone. Over 75000 people will be evicted. Additional displacement is likely to be caused during social and environmental rehabilitation work undertaken to repair the dislocation and damages caused by the project.

18. Narmada Valley Project (NVP). The world's largest river valley project has attracted the greatest attention. The 30 big dams and over 3000 medium and minor dams are envisaged at cost of Rs. 25,000 crores. It would displace one million people, mostly tribals, submerge 56,000 ha of fertile agriculture land. Total forest area of nearly 60,000 ha. will be destroyed . About 25 species of birds will be deprived of their habitats.
19. Bodhghat Project. On Indravati river in M.P. is in heavily forested Bastar district. The project will destroy teak and sal forests. The criticism of the project forced the Government and the World Bank to reconsider it.

Man under Effect: MAB in fact is the outcome of the experience of those involved in the International Biological Programme (IBP). It was realised that several problems require collaboration of natural and social scientists, planners and managers and the local people. MAB was conceived at the International Biosphere Conference at its 16th session in 1970. The programme was formally launched by UNESCO in November 1971, when the MAB International Coordinating Council held its first session and identified 13 project areas of cooperative research. One more area of project area was added in 1974.

10

GLOBAL SCENE

Environment is defined as "the sum total of all conditions and influences that affected the development and life of organisms." The interplay of material cycles and energy flow in natural ecosystem generates a self-correcting homoeostasis with no outside control or set points. The ecosystem is capable of self-maintenance. However, the equilibrium is very sensitive to external stimuli such as human activities promoted by socio-economic goals.

The global community recognised the seriousness of the U.N. declared 1990s as the International Decade for "Natural Disaster Reduction." As a part of this world wide programme of education and action has called upon to make people aware of the danger ahead and the steps they must take to check and mitigate their ill effects. As we know that Global climate is changing due to green house effect" *i.e.* due to the accumulation of green house gases such as carbon dioxide, chloro-fluorocarbons (CFCs) methane, nitrous oxide and ground level ozone gas.

Half of the global warming is alone by carbondioxide gas.

Temporary stability can be achieved if annual emission of CO, could be reduced by shifting fossil fuel use from coal to oil to natural gas. This can also be achieved if we shift to other alternate sources of energy like solar, geothermal, tidal, hydro etc.

The ozone layer present in stratosphere protects us from the ultraviolet rays which would cause different ailments like sunburns, skin cancer, eye damage etc. This is also depleting because of released of CFCs, nitrous oxide and other gases. In order to avert the worst effect of warming due to green house effect serious efforts are being made all over the world to reduce the release of these gases into the atmosphere.

Global Warming

Global warming or green house effect is the warming of the earth due to emission of the harmful gases from the earth, which forms the blanket over the earth and do not allow the gases to move upward to the atmosphere. Since CO_2 is confined exclusively to the troposphere, its higher concentration may act a serious pollutant. Under normal conditions (with normal CO_2 concentration).

The temperature at the surface of the earth is maintained by energy balance of the sun rays that strike the planet and heat that is radiated back into the space. However, when there is an increase in CO_2 concentration, the thick layer of this gas prevents the heat from being re-radiated out. This thick carbon dioxide layer thus functions like the glass panels of a green house (or the glass windows of a motor car), allowing the sunlight to filter through but preventing the heat from being re-radiated out in outer space.

This is the so called green house effect. Thus most heat is absorbed by CO_2 layer and water vapours in the atmosphere, which adds to the heat that is already present. The net result is the heating up of earth's atmosphere. Thus increasing CO_2 levels tend to warm the air in lower layers of atmosphere on a global scale. Other gases of the green house effect on earth are methane, chlorofluorocarbons and nitrogen oxides. These are called green house gases. These gases are emitted to the atmosphere through burning of fossil fuels like coal, wood, oils etc.

It is estimated that more than 18 x 1012 tonnes of CO_2 is being produced annually from the fossil fuels only CO_2 concentration has increased from 280 ppm in 1800 to 359 ppm in 1994 due to increasing use of fossil fuels and decreasing forest cover. More methane is being released from paddy fields due to intensive cultivation, cattle due to large population and escape from gas plants. Combustion of fossil fuels (along with excessive use of nitrogen fertilizers) is also the causative agent for higher concentration of nitrogen oxides.

They react with hydrocarbons to form ozone. As there is an increase in the release of CO_2 it is estimated that temperature increases after every 100 years from 0.3° to 0.7°C. If these gases are not minimised in the year 2030 the concentration of CO_2 will increase twice the present range and the temperature of the earth surface will increase by 1.5°C to 4.5°C.

Effects

1. Global warming shall cause partial melting of polar and alpine ice caps that would result in raising sea level. Already the sea level has risen 15 cm in the past century. If the trend continues, there is a danger of submersion of large area *e.g.* Maldives, and six other Coral atoll countries, several thousand islands.
2. A continuous increase of green house gases, leads to the rise in mean global temperature which affects the climate leading to change in rainfall pattern, conversion of fertile lands into deserts, shortage of water due to evaporation.
3. Due to change in weather conditions, the production of crops is also affected and reduces the yield of grains like rice, maize, wheat etc.

4. Chances of hurricanes, cyclones and floods will increase.
5. Green house effects disturb the food chain and destroy many cold habitat species.
6. The forests present in the middle latitudes will be wiped out.
7. Several lakes would dry up.

So, UNEP has appropriately chosen the slogan "Global Warming: Global Warming" and since 1989 5th June is celebrated as World Environment Day.

These problems can be overcome by awareness among people and plantation of trees and use of non-conventional sources of energy.

Issues of Conservation

Conservation has been defined as "the management for the benefit of all life including human kind of the biosphere so that it may yield sustainable benefit to the present generation while maintaining its potential to meet the needs and aspiration of the future generation." As the earth is the only planet where life can exist, population on the earth is increasing at an alarming rate which is putting stress on the natural resources at much larger scale. Continuous increase in population caused an increasing demand for resources. This created a situation when the non-renewable resources may come to an end after some time.

Basically there are two main reasons for exploitation of environment and its resources. One reason is poverty while the other is rich people. Although development is necessary for man but it should be in a way that it should not destroy the resources. All the available resources available on the earth should be used judiciously so that they should not be depleted. The people should be made aware of the consequences of this depletion. Since the natural resources are over exploiting there should be a check or control over it.

Purposes and Objectives

The main objectives to conserve the natural resources are:

(i) to maintain the proper ecological balance and life support system.

(ii) to preserve biological diversity.

(iii) to ensure that any utilisation of species and ecosystems is sustainable.

Conservation, therefore, makes important contributions to social and economic development.

Making of Plans: At global level there are only two strategies

(a) Special Interest Conservation Strategies. Conservation is to be done of only important and limited resources.

(b) Total Ecosystem Conservation. Conservation of all species of the organisms on the earth.

U.N. Conference on Human Environment for Stockholm declaration in year 1972 from 5th June to 16th June held in Stockholm. The following recommendations were made by the conference.

(i) Every human being has the right to live in healthy environment and his fundamental duty is to protect the environment and its surroundings.

(ii) Ecological balance should be maintained.

(iii) For the protection of the society the wild life should be properly managed.

(iv) Natural renewable resources should be maintained on the earth.

(v) No interference in geological balance, or over exploitation of natural resource as they lead to pollution.

(vi) Waste and pollutants should not be discharged directly into oceans.

(vii) Science and technology should be used for the identification of various disasters of environmental pollutants.

(viii) Every country must take into account that any work done in their country should not effect the other country.

(ix) Environment effects the economic development and quality of living of the human beings.

(x) Every one should be given education about the ill effect of deterioration of environment.

The late Prime Minister, Smt. Indira Gandhi played a vital role in the above conference. She had taken initiative and come forward with a definite programme on environmental conservation. She made 42nd amendment in the constitution. According to this amendment, the problems regarding forests, wild life and environment were considered. Laws have been made and fundamental duties imposed on every citizen of this country.

Under the auspices of the International Union of Conservation of Nature and Natural Resources (IUCN), World Wildlife Fund (WWF) and UNEP, a World Conservation strategy was prepared and released for adoption and implementation in India on March 5,1980. The WCS is to ensure the management of human use of biosphere in a manner which may yield greater sustainable benefit to the present generation while maintaining its potential to meet the needs and aspirations of future generation.

Keeping in view the need for environmental protection, the Indian Parliament passed the water (Prevention and Control of Pollution) Act, 1974 which became effective from 23 March 1974. The Central Board for Prevention and Control of Water Pollution (CBPCW) was constituted in Sept. 1974. The Air (Prevention and Control of Pollution) Act 1981 was promulgated on 16 May, 1981. The Environment (Protection) Act 1986, was promulgated on 23 May, 1986.

The main instruments presently for control of air, water and noise pollution in country are:

(1) The water (Prevention and Control of Pollution) Act 1974 .

(2) The Air (Prevention and Control of Pollution) Act 1981.

(3) The Environment (Protection) Act 1986.

(4) The Motor Vehicles Act, 1988.

Besides there are other acts to conserve our environment.

Factories Act (1948) — Amended in (1987): In Section 12 of our constitution it is stated that it is the duty of every industrialist to properly treat and manage the waste according to the plans laid down by the state government.

(i) In Section 11 of the act, it is given that it is the duty of the factory owner to check the factory is free from any effluent.

(ii) Any effluent producing factory should be away from human habitat.

(iii) Only limited amount of harmful waste should be emitted.

(iv) Machinery should be properly maintained and should be periodically checked.

Environment (Conservation) Act 1980 — (Amended 1988): According to this law:

1. Natural forests on the earth cannot be changed into other kind of plantation without prior permission of the Government.
2. If any one converted forest area for development project, the same area is to be planted by him.
3. Forest planning and management should be stressed.
4. No area should be deforested with the purpose of reforesting it.
5. There should be controlled grazing.

6. A forestation near the hills and slopes of the hills.
7. Encouragement to the tribal community.
8. Penalties for disobeying laws.

Forest (Conservation) Act, 1980—(Amended 1988): National Forest Policy 1952 enunciated that one third of the geographic area of the country should be under forests. However there had been continuous deforestation in the country for various reasons, and it is estimated that 4.23 8 Mha of forest land was officially diverted to non-forest purposes between 1951-52 and 1979-80. With a view to conserve forests, Govt. of India could enact the Forest (Conservation) Act, 1980.

1. Act was enacted with a view to check indiscriminate dereservation and diversion of forest land to non-forest purposes.
2. Under this act prior approval of Central Government is required before any reserved forest is declared as dereserved, or forest land is diverted to non-forest purposes.
3. If diversion is permitted, compensatory afforestation is insisted upon and other suitable conditions imposed.
4. Where non-forest lands are available compensatory afforestation be raised over equivalent area of non-forest land.
5. Where non-forest lands are not available, compensatory plantation be raised over degraded forests twice in extent to the area being diverted.
6. A control should be exerted over shifting cultivation and encroachments.
7. Grazing problems of the area should be studied and appropriate measure be adopted.
8. All forest working plans should stress conservation and have multi-disciplinary approach.
9. All critical areas in the hills, catchment areas, steep slopes and other parts under erosion must be protected and quickly afforested.

Six regional offices have been set up to monitor the conditions of forests and to find out the steps for their conservation. There are located at Bangalore, Shillong, Bhopal, Bhubaneshwar, Lucknow and Chandigarh.

However there are certain limitation to this Act.

(i) Since the prices of timber at rise therefore stealing and smuggling of finest timbers have increased.

(ii) The people have started encroachment in the forest area.

(iii) Since the number of grazing animals is on increase they usually roam in restricted areas.

(iv) The forests have been destructed for construction of roads, water supply schemes etc.

(v) It is very difficult in forests to apprehend the offenders.

Wildlife (Protection) Act, 1972—(Amended 1991): Wild life includes both plants and animals, is a living component of nature. It is must to maintain ecological balance, to prevent soil erosion and to obtain products which are economically important such as wood, medicines and drugs, resins, gums, fodder etc. A number of wild life acts have been made from time to time, by state as well as Union Government for wildlife conservation such as: (a) Madras Wild Elephant Preservation Act, 1873 (b) All-India Elephant Preservation Act, 1879 (c) The Wild Birds and Animals Protection Act, 1912 (d) Bengal Rhinoceros Preservation Act, 1932. (e) Assam Rhinoceros Preservation Act, 1954 (f) Wild Life (Protection) Act, 1972.

Indian Board of Wildlife (IBWL) is the main advisory body of the Govt. of India. It was first constituted in 1952 as an advisory body under the name Central Board of Wildlife. Later it was renamed as IBWL. At its first meeting, the board made a recommendation for unified legislation for wildlife conservation in India.

The Wildlife (Protection) Act was enacted in 1972 which was amended in 1991, which has been adopted by all states.

The Indian Board of Wildlife which act under this act is chaired by Prime Minister. IBWL has asked both B.S.I. (Botanical Survey of India and Z.S.I. (Zoological Survey of India) to prepare a list of threatened species (which is likely to become extinct) of both plants and animals. These species are recorded in Red Data Book of IUCN (International Union for Conservation of Nature and Natural Resources).

At the 15th meeting of the IBWL held on 1st Oct. 1982, the then Prime Minister, Late Smt. Indira Gandhi gave a 12 Point strategy for an Action Plan for the conservation of wildlife in India. This included the establishment of a net work of scientifically managed protected areas including National Parks, Sanctuaries, Biosphere reserves. National Parks are the areas maintained by Govt. where cultivation, grazing, forestry and habitat manipulations are not allowed. Sanctuaries are tracts of land with or without cakes where hunting of animals is not allowed but other activities such as tilling of land, collection of forest products etc. are allowed.

Biosphere reserves are multipurpose protected areas where wildlife, tribals, domesticated plants and animals are allowed to live in harmony.

The main recommendations of the act are:

(i) Trapping and hunting of wild life should be prohibited.

(ii) Conservation and effective control of poaching wildlife.

(iii) Setting of national parks, sanctuaries and zoological gardens.

(iv) Ban on the export of living animals, skins, fur, leather and other wild life products.

(v) Human activity is not allowed in the core zone of biosphere reserve and national parks.

(vi) Uprooting, picking and sale of endangered plants not allowed.

(vii) No arm licence is to be issued in an area within 10 km of a Sanctuary.

Limitations

(a) Less attention is given to endangered plant species

(b) Exploitation of plant species continue under the garb of traditional medicine as the plants of medicinal importance are not cultivated.

(c) Wildlife authorities take a lot of time to decide whether an animal is dangerous and by the time it is decided it would have killed a number of domestic animals.

(d) Stress is given only on conservation of one or two species not all.

Environment (Protection) Act, 1986: The Act was promulgated to provide for the protection and improvement of environment and matters connected therewith. As per Environment (Protection) Act 'environment' includes water air and land and the inter relationships which exists among and between water, air and land and human beings, other living creatures, plants, micro-organisms and property; "environment pollutant" means any solid liquid or gaseous substance present in such concentration as may be injurious to environment; "environment pollution" means the presence in the environment of any environmental pollutant, and "hazardous substance" means any substance which is liable to cause harm to human beings and other living creatures. The Act provides general power to the central government to take all necessary measures for the purpose of:

(a) Protecting and improving the quality of the environment.

(b) Preventing, controlling and abating environmental pollution.

The functions of Central Govt., are:

1. Planning and execution of nation wide programme for the prevention, control and abatement of environmental pollution.

2. Laying down standards for quality of environment in its various aspects.
3. Laying down standard for emission or discharge of environmental pollutants from various sources.
4. Carry out and sponsor investigations and research.
5. Collection and dissemination of information on environmental pollution.
6. Examination of such manufacturing processes materials and substances which are likely to cause environmental pollution.
7. Demarcation of areas in which certain industries/processes/operations shall be not or shall be carried out subject to certain safeguards.
8. Laying down procedures and safeguards for tackling of restricted prohibited and hazardous substances.
9. Preparation of manuals, codes and guides regarding prevention, control and abatement of environmental pollution.
10. Establishment or recognition of environmental laboratories.

Air (Prevention and Control of Pollution) Act, 1981 (Amended 1987): Air (Prevention and Control of Pollution) Act was promulgated under article 253 of constitution for prevention control and abatement of air pollution by creating central and state boards.

Central Board exercises powers and performs the functions of State Board for Union Territories either directly or by delegating powers.

The various functions of Central Board are:

(i) To provide guidance and technical assistance to state boards and the industries.

(ii) To advise both Central Government and State Government regarding improvement in methods related to check air pollution.

(iii) To provide training to the persons involved in the field of air pollution.

(iv) To set up laboratories to check all kinds of samples.

(v) To educate people with the assistance of mass media.

The various functions of State Board are:

(i) To advise State Government to combat with the problem of air pollution.

(ii) To collect information regarding causes, prevention and control of air pollution.

(iii) To lay down standards for air quality.

(iv) To inspect air quality periodically to check air pollution.

As per Air (Prevention and Control of Pollution) Act both Central and State Boards have been given certain powers to meet the consequences due to air pollution.

(a) To declare any area within the state as air polluting area.

(b) Before setting up any unit every industrial establishment has to take clearance from the board. The board has got power to refuse or grant consent depending upon whether the industry fulfils the specification laid down by the board.

(c) The board has got power to stop industrial operations in air pollution control areas.

(d) The board collects the samples of air or emission for analysis.

(e) The board has got power to cancel the consent given to industry at any time.

(f) The board officials have got power to go to any industry at any time to check compliance with requirements of act.

(g) The board has got power to prosecute any defaulter.

The Air (Prevention and Control of Pollution) Act, 1981 was amended in 1987 to remove the difficulties encountered during implementation, to confer more powers on the implementing agencies and to impose more stringent penalties for violation of the provisions of the Act. The main point was also to amend the definition of air pollutants to include noise also.

The Water (Prevention and Control of Pollution) Act, 1974 (Amended 1988): The Water (Prevention and Control of Pollution) Act was enacted under article 252 (1) of Constitution as measure to prevent and control water pollution.

The pollution under the act includes (a) any contamination in water (b) any physical, chemical or biological alteration in water (c) direct or indirect discharge of sewage, trade effluent or any other substance in water, (d) the discharge causing nuisance or harmful to public health. Central and State boards were created to meet the problem regarding water pollution.

The water (Prevention and Control of Pollution) Act, 1974 was amended in 1988. An important amendment was to rename the Central State Boards for Prevention and Control of Water Pollution as Central/ State Pollution Control Boards as Boards also deal with air pollution. More powers were given to CPCB.

Functions of Central Board

(i) To advise Central Government and State Governments about issues related to water pollution.

(ii) To provide guidance and training to persons in the field of water pollution.

(iii) To educate people through mass media.

(iv) To provide technical assistance and guidance to state boards and industries.

(v) To set up laboratories to analyse samples.

Functions of State Board

(i) To advise state Government regarding issues related to water pollution.

(ii) To seek guidance and training of persons connected with prevention and control of water pollution.

(iii) To collect information regarding causes, prevention and control of water pollution.

(iv) To organise programmes to control water pollution.

(v) To find out recent methods for disposal of treated sewage and effluents.

Water (Prevention and Control of Pollution) Act has provided different powers to both Central/State Pollution Control Boards:

1. It is the duty of each industrial establishment to take consent from state board on a specific application about the methods of treatment of disposal of sewage and other industrial effluent be discharged in the water body.
2. Board has got power to take sample of any sewage, industrial effluent from any water body.
3. Board has power to enter any industrial premises for inspection.
4. Board has got power to move any application to court after an offence is detected.
5. Board can ask closure of any unit any time It can stop water supply, electricity to the units of defaulters.

The Motor Vehicle Act, 1938 (Amended 1988): This Act has come into force from 1st July 1989. The salient features of the provisions of the Act, besides other things also include the requirements to be observed in matter of vehicle fitness. Some of the features are as follows:

1. For registration of a new vehicle, the same is to be produced before the registering authority for inspection to satisfy that vehicle is fit for registration. The certificate of fitness is to be issued for a new vehicle for a period of two years. Thereafter, the renewal shall be for period of one year till the vehicle attains the age of 10 years.
2. Every vehicle is required to meet the safety stands of components.

3. Each vehicle must have all the components of the standard laid down by Bureau of Indian Standards.
4. The vehicle must have tested by VRDE, Ahmadnagar, ARAI, Pune or CMITI, Budni or any other agency specified by the Central Government.
5. Manufacturer is to ensure that the vehicle does not cause pollution *i.e.* It is pollution free.
6. The horn to be used is to be in accordance with the approved specifications of Bureau of Indian Standards.
7. The vehicle should be fitted with tune up and catalytic converter.
8. Each vehicle is required to obtain "Pollution under control" certificate every 6 month.
9. All future vehicles are to have engines based on unleaded petrol.
10. Transportation of Goods of dangerous or hazardous nature to human life has been regulated under the Act. The vehicle carrying such goods should have prescribed labels indicating the nature of hazardous goods. The package containing such goods should also bear specific labels.

Bibliography

Abbasi, S.A. : *Wetlands of India : Ecology and Threats,* Discovery Publishing House, New Delhi, 1997.

Adkins, R.M. : *Annual Review of Ecology and Systematics,* A. R., London, 1993.

Akhtar, R. : *Disease Ecology and Health,* Rawat Publications, Jaipur, 2001.

Alison, J. : *Spatial Ecological : Economic Analysis for Wetland Management,* Cambridge, London, 2004.

Anthony, J. : *Handbook of Ecological Restoration,* Cambridge, London, 2002.

Antonovics, Janis : *Integrating Ecology and Evolution in a Spatial Context,* Cambridge, London, 2001.

Banerjee, B.N. : *Environmental Pollution and Bhopal Killings,* Gyan Books, New Delhi, 1987.

Barker, G. : *Physiological Plant Ecology,* Cambridge, London, 2002.

Berger, W.H. : *Abrupt Climatic Change,* D. Radiel Dordrecht, Boston, 1990.

Bhargava, Gopal : *Ecological Politics : Different Dimensions,* Gyan Books, New Delhi, 2002.

——— : *Marine Ecosystems,* Gyan Books, New Delhi, 2001.

Bhusan, Bibhuti : *Social Ecology of Forest Resources,* Gyan Books, New Delhi, 2004.

Billihgs, W.D. : *Plant, Man and the Ecosystem,* MacMillan, London, 1971.

Bookchin, M. : *The Philosophy of Social Ecology,* Rawat Publications, Jaipur, 1996.

Chandra, Ramesh and Aneja, Ritu : *Corporate Governance for Sustainable Environment,* Gyan Books, New Delhi, 2004.

Colin, J. : *Issues in Environmental Economics,* Blackwell, London, 2002.

Crawford, T.J. and Hewitt, G.M. : *Genes in Ecology,* Cambridge, London, 1992.

Crosby, A.W. : *Ecological Imperialism : The Biological Expansion of Europe,* Cambridge, London, 2000.

Dakshinamurthi, C. : *Water Resources of India and their Utilization in Agriculture,* Oxford University Press, New Delhi, 1973.

Dass, Sujata K. : *Environment Challenges and Sustainable Future,* Gyan Books, New Delhi, 2003.

Dasseman, R.F. : *Environmental Conservation,* John Wiley, New York, 1976.

Dave, M. : *Urban Ecology and Levels of Development,* Rawat Publications, Jaipur,

David, Khemchand : *Advanced Dictionary of Ecology and Environment,* Dominant Publishers, New Delhi, 2000.

DeVere, Burton : *Ecology of Fish and Wildlife,* Delmar, London, 1996.

Dickinson, L. : *Ecology and Evolution of Cooperative Breeding in Birds,* Cambridge, London, 2004.

Dodson, S. I. : *Ecology,* Oxford, New York, 1998.

Dutta, Ashish : *Biodiversity and Ecosystem Conservation,* Gyan Books, New Delhi, 2001.

Dwivedi, S.D. : *Naturopathy : For Perfect Health,* Gyan Books, New Delhi, 2002.

Eric, R. Pianka : *Evolutionary Ecology,* Harper Collins, London, 1994.

Fritz, L. Knopf : *Ecology and Conservation of Great Plains Vertebrates*, Springer, London, 1997.

Gaan, Narottam : *Environmental Security Concept and Dimensons*, Gyan Books, New Delhi, 2004.

Gary, A. Klee : *The Coastal Environment : Toward Integrated Coastal and Marine Sanctuary Management*, Prentice Hall, London, 1999.

George, A. Maul : *Climatic Change in the Intra : Americas Sea*, Arnold, New Jersey, 1993.

Gerry Closs, Barbare Downes and Andrew, Boulton : *Freshwater Ecology : A Scientific Introduction*, Blackwell, London, 2004.

Gordon, D. : *World Population Problems*, Park Press, Campaign, 1984.

Grafton, Q. : *Economics of the Environment*, Blackwell, London, 2004.

Gupta, K. : *Energy and Environment in India : A Study of Energy Management*, Gyan Books, New Delhi, 2002.

Hourani, G.F. : *Arab Seafaring in the Indian Ocean and Early Medieval Times*, Princeton University Press, Princeton, 1951.

Husain, M. : *Evolution of Geographical Thought*, Rawat, Jaipur, 1995.

Hussen, A. : *Principles of Environmental Economics*, Routledge, New York, 2004.

Hutchings, J. : *The Ecological Consequences of Environmental Heterogeneity*, Cambridge, London, 2000.

Ian, D. Whyte : *Climating Change and Human Society*, Arnold, New Jersey, 1995.

Ian, F. Spellerberg : *Monitoring Ecological Change*, Cambridge, London, 2005.

James, K. Lein : *Integrated Environmental Planning*, Blackwell, London, 2003.

James, M. Bullock : *Dispersal Ecology*, Cambridge, London, 2002.

John, E.FA : *Evolution and Ecology of Macque Societies*, Cambridge, London, 1996.

John, N. Rayner : *Dynamic Climatology : Basis in Mathematics and Physics,* Blackwell, London, 2001.

Joshi, M.V. : *Theories and Approaches of Environmental Economics,* Atlantic Publishers, New Delhi, 2004.

Kaur, Karanjot : *Green Revolution : Ecological Implications,* Dominant Publishers, New Delhi, 1999.

Khanna, D.R. : *Microbial Ecology : A Study of River Ganga,* Discovery Publishing House, New Delhi, 2004.

Kumar, Chattopadhyay : *The Desperate Delta : Social Ecology of Sunderbans,* Gyan Books, New Delhi, 1995.

Kumar, Shashi : *Human Ecology for Globalization : Human Ecology in Action,* Atlantic Publishers, New Delhi, 2003.

——— : *Human Ecology for Globalization : New Dimensions for Economic Growth,* Atlantic Publishers, New Delhi, 2004.

Madan, Mohan : *Ecology and Development,* Rawat Publications, Jaipur, 2000.

Martin, T.E. : *Ecology and Management of Neotropical Migratory Birds,* Oxford, New York, 1995.

Mash, G. P. : *Man and Nature Physical Geography as Modified by Human Action,* Charles Scriber, New York, 1964.

Max, N. : *The Environmental Revolution,* McGraw-Hill, New York, 1970.

Michael, Moss : *Issues and Perspectives in Landscape Ecology,* Cambridge, London, 2005.

Moore, Francis : *Environment and Society,* Dominant Publishers, New Delhi, 2003.

Mukherjee, Roma : *Environmental Management and Awareness Issues,* Sterling Publishers, New Delhi, 2002.

Muthiah, S. : *An Atlas of India,* Oxford University Press, New York, 1992.

Myers, H. : *Ecology and Control of Introduced Plants,* Cambridge, London, 2003.

Nadarajah, M. : *Culture, Gender and Ecology : Beyond Workerism*, Rawat Publications, Jaipur, 1999.

Nigel, J.R. : *Mountains at Risk : Current Issues in Environmental Studies*, Manohar Publishers, New Delhi, 1995.

Nigel, Webb : *Ecological Reviews*, Cambridge, London, 2005.

Overbeck, J. : *Microbial Ecology of Lake Plubsee: Ecological Studies*, Springer, London, 1994.

Pandey, Mahendra : *Environment Pollutants & Women's Health*, Dominant Publishers, New Delhi, 2003.

Paul, A. : *Introduction to Ecology*, John Wiley and Sons, New York, 1973.

Peter, J. Grubb and John, B. Whittaker : *Toward a More Exact Ecology*, Cambridge, London, 1989.

Puri, G. : *Essential Soil Science : A Clear and Concise Introduction to Soil Science*, Blackwell, London, 2002.

Rampinot, M.R. : *Climate: History, Periodicity and Predictability*, Von Nostrand Reinholt, New York, 1987.

Ravi, V.S. : *The Living Adventure : Nature and Biology*, Manohar Publishers, New Delhi, 2001.

Raza, M. : *Development and Ecology*, Rawat Publications, Jaipur, 2000.

Ring, T. Carde and Jocelyn, G. Millar : *Advances in Insect Chemical Ecology*, Cambridge, London, 2004.

Rob, H.G. Jongman : *Ecological Networks and Greenways*, Cambridge, London, 2004.

Sarah, J. Woodin and Mick, Marquiss : *Ecology of Arctic Environments*, Cambridge, London, 1997.

Sheth, P. : *Environmentalism : Politics, Ecology and Development*, Rawat Publications, Jaipur, 1997.

Shukla, S.K. and Srivastava, P.R. : *Ecology and The Environment*, Commonwealth Publishers, New Delhi, 1992.

Sinha, R.K. : *Eco-Politics and Development in India*, Commonwealth Publishers, New Delhi, 1996.

Stamou, G.P. : *Arthropods of Mediaterranean : Type Ecosystems,* Springer, London, 1998.

Thomas, R. : *Man's Impact on the Environment,* McGraw-Hill, New York, 1988.

Thomson, T. : *Introductory Essay to Flora Indica,* MacMillan, London, 1855.

Trivedi, P.R. and Dudarshan, K.N. : *India Ecology and Environment,* Commonwealth Publishers, New Delhi, 1994.

Usher, Michael : *Ecology, Biodiversity, and Conservation,* Cambridge, London, 2005.

Vayda, A.P. G. : *Man in the Pacific Islands: Essays on Geographical Change in the Pacific,* Oxford University Press, New York, 1972.

Venus, W. : *Kanpur City: A Study in Environmental Pollution,* Tara Book Agency, Varanasi, 1982.

Vogt, K.A. : *Ecosystems : Balancing Science with Management,* Springer, London, 1997.

Walter, G.H. : *Insect Pest Management and Ecological Research,* Cambridge, London, 2003.

Wilfried, B. : *Atmospheric Pollution,* McGraw Hill, New York, 1972.

William, W. : *Integrating Landscape Ecology into Natural Resource Management,* Cambridge, London, 2002.

Yadav, P.R. : *Environmental Ecology,* Discovery Publishing House, New Delhi, 2003.

Yeates, M. : *Plant Ecology,* Harper & Row, New York, 1980.

Zelinsky, W. : *A Study of Ecology,* Oxford University Press, New York, 1970.

Zuckerman, Ben : *Human Population and the Environmental Crisis,* Jones & Bart., New Jersey, 1996.

Index

A

B

C

D

E

F

G

H

I

J

K

L

M

N

O

P

Q

R

S

❑❑❑